D1065355

MARITIME RADICAL

Fillmore Family Tree

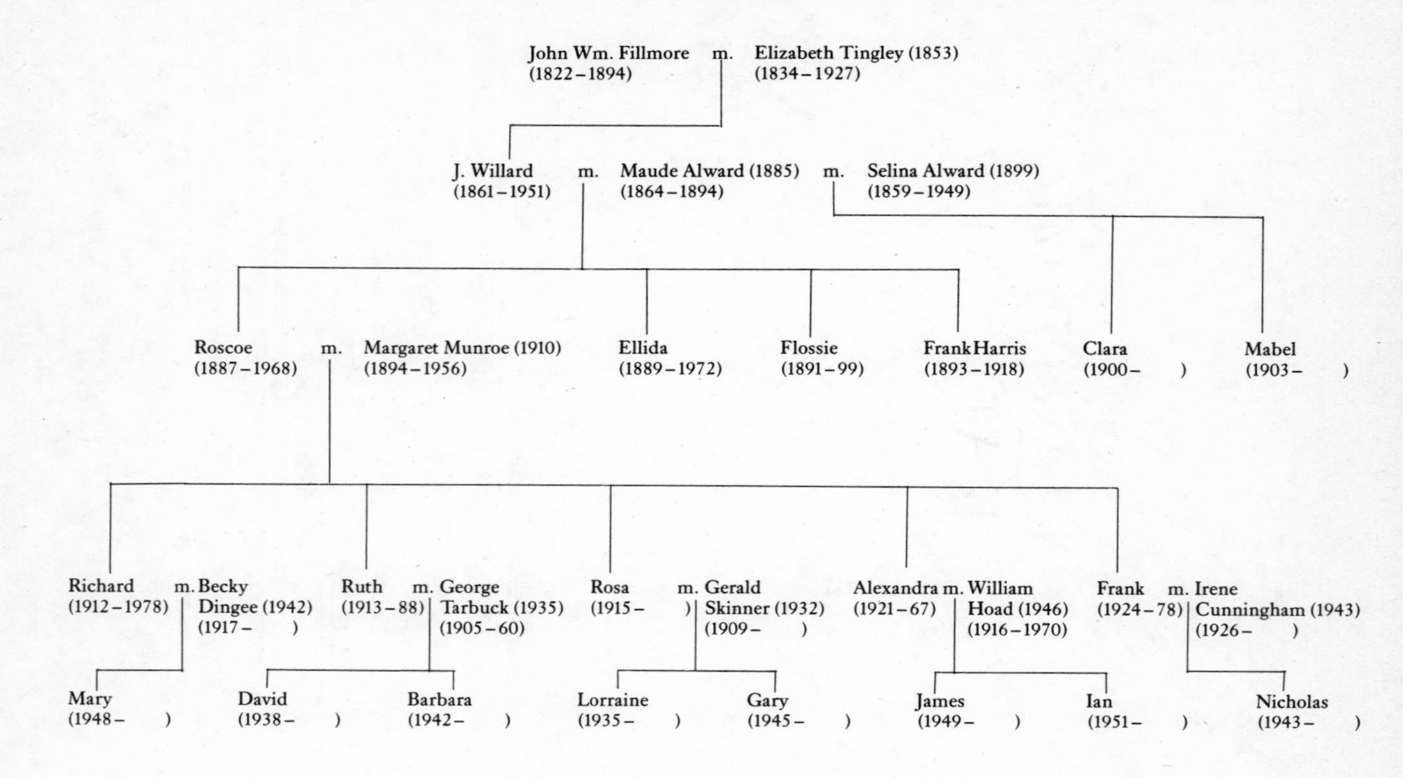

MARITIME RADICAL

The Life & Times of Roscoe Fillmore

NICHOLAS FILLMORE

between the lines

Published by Between The Lines
394 Euclid Avenue, #203
Toronto, Ontario
M6G 2S9
Canada

Cover design by Goodness Graphics, Toronto
Typeset by Techni Process, Toronto
Printed in Canada

Map on pg. xii from *The Canadian Encyclopedia* by Hurtig Publishers. Used by permission of the Canadian Publishers, McClelland & Stewart, Toronto, and Energy, Mines and Resources.

Between The Lines receives financial assistance from the Canada Council, the Ontario Arts Council, and the Ontario Ministry of Culture and Communications, through the Ontario Publishing Centre.

Front cover photo: Roscoe returning to Canada from Siberia, 1923.

Canadian Cataloguing in Publication Data
Fillmore, Nicholas, 1943-
Maritime Radical

Includes index.
ISBN 0-921284-49-7 (bound) ISBN 0-921284-50-0 (pbk.)

1. Fillmore, Roscoe A. 2. Socialism—Canada—History—20th century. 3. Canada—Politics and government—20th century. 4. Socialists—Canada—Biography. 5. Horticulturists—Canada—Biography. I. Title.

HX104.7.F55F55 1992 320.5'31'092 C92-093719-5

Contents

Acknowledgements

IN THE FOUR YEARS I spent on this project I received the help and support of my family and many friends and colleagues. I want to thank the many people who have assisted me in piecing together Roscoe Fillmore's life, in particular Roscoe's daughter Rosa Skinner, and his long-time comrade Dane Parker.

My special thanks to editor Robert Clarke, a strong supporter of this book from the beginning, to whom I owe a debt of gratitude for substantially improving the finished product. My thanks also go to my publisher, Between The Lines.

From my family, I am especially grateful to my mother, Irene Fillmore, for reading the manuscript and assisting with my research. Thanks also to Clara Reid, Mabel Gross, Mary Dingee Fillmore, Barbara Tarbuck, Becky Fillmore, Lorraine Skinner, and Harold Milton.

My thanks for providing critical analysis of the manuscript to Jamie Swift and Ian McKay, who provided special encouragement. I also am grateful to several historians who went out of their way to assist me, including Mike Earle, David Frank, Nolan Reilly (who was instrumental in having Roscoe's papers deposited at Dalhousie University Archives), Greg Kealey, Barry Grant, Cheryl Lean, Graham Metson, John Bell, and John Shuh.

A special thanks to Charles Armour of the Dalhousie University Archives and to the many other archives and libraries that helped me with my research. While I found the staff of these institutions to be extremely helpful, I am concerned that budget cuts have put extreme pressure on many staff members and limited their ability to adequately assemble and catalogue material so that books such as mine can be written. They deserve better financial support.

For submitting to interviews and providing information during this project I am grateful to Kell Antoft, Mildred Ashby, Vern Bigelow, Gary Burrill, Don Craig, Fran Fassett, Elizabeth Etter Fillmore, Marj Hancock, Anne Hutten, Ruth Marvin, Kaye Murray, George MacEachern, Ethel Meade, Ina Milburn, "Scotty" and Alice Munro, Harold Porter, J.B. Salsburg, Thelma (Martin) Scott, and Ken Wilson.

My special thanks to the Office of the Information Commissioner of Canada for assistance in getting as many pages as possible of Roscoe's RCMP records from the Canadian Security Intelligence Service. I also wish to thank the Canada Council's Explorations Program and Program Officer Megan Williams who provided the seed money to make the book possible, and the Ontario Arts Council, which provided additional funding through its Writers' Reserve program.

Finally, I am greatly indebted to Barbara Tessman — who became my wife while I was working on this project — for her many hours of editing, help with rewriting parts of the manuscript, as well as advice and encouragement provided over the past five years.

N.F., November 1991

Preface

ROSCOE FILLMORE WAS my grandfather. When I was growing up in the Annapolis Valley of Nova Scotia, in the early 1950s, he and my father, Frank, operated the family-owned Fillmore's Valley Nurseries. I remember how proud I was when Bamp—as I called him—became well known for the garden books he was writing.

As a small child in summer I accompanied my grandfather to Grand Pre Memorial Park, where he was in charge of the gardens. I remember happy, sunny days at the park when he would let me open the door to the little house where the ducks were kept and I would chase them out into the nearby ponds, which were surrounded by the tall willow trees my grandfather loved so dearly.

By the age of thirteen I was working part-time at the nursery, and my grandfather and father were showing me the ropes of the family business. I loved the smell of the damp greenhouse in winter, the budding and breaking out of the first leaves of spring, and I grew to understand why four generations of Fillmores had been in horticulture. Roscoe dedicated his book *Green Thumbs* to his eight grandchildren, and in my copy he wrote a note:

> Dear Nicky,
>
> Do you remember when you climbed on my back when I was planting pansies? A few days ago I saw you hustling pansies out of the field for a customer. I think you could be a good nurseryman some day and I trust this book may help you and that you may learn far more about plants than 'Bamp'

ever knew. To learn about things is what we are here for. Remember, we want there to always be a Fillmore Nursery.

Grandpa,
June 20, 1953

My grandfather not only taught me about horticulture but also shared his ideas about politics and his tremendous knowledge about the ways of the world. When I had homework and needed an explanation about some point in history or geography, I could go to Grandpa—he always took time to talk—and I was amazed that he could tell me very human stories to explain why certain things had happened. But it was puzzling, too: My grandfather's version of events was often quite different from what I was told in school. In fact more than once I got into difficulty with my teachers when I presented them with what Bamp had said.

When I was growing up my grandfather seldom talked much about his past. He occasionally mentioned having been in Russia or the visits of Communist Party leader Tim Buck to Centreville, but he seldom provided details. I knew he had been a communist, and this was driven home to me in a painful way by my schoolmates when I began to attend Kings County Academy in Kentville. In Grade 8 I opened one of my school books and read "Commie" scrawled on the first page in dark blue ink. The word was accompanied by a large Nazi swastika—I realized only later that the person who had defaced my book wasn't too clear about his history. What I did realize at the time was that I was being singled out as different from everyone else. This sort of thing continued to happen from time to time during the years I attended the school.

After the early 1960s, when my family moved from the Valley to Dartmouth, Nova Scotia, I had less contact with my grandfather. A couple of years later I went off to my first job as a reporter at *The Guardian* in Charlottetown. What I discovered in this job—somewhat to my surprise—was that I had been instilled with social and political values quite different from those of most of the other reporters. I found I had a keen sense of social justice and a soft spot for the underdog: a direct legacy, I believe, from Roscoe. It seemed to me that my journalism benefited from the knowledge and moral code he had helped to implant in me.

After Roscoe Fillmore's death, in 1968, his personal papers and parts of an unpublished autobiography went to the Dalhousie University Archives in Halifax; and it wasn't until nearly twenty years later that I became interested in his life as a matter of historical importance. A group of socialists in the Maritimes had made a tradition of celebrating his birthday—July 10—with the Roscoe Fillmore Memorial Picnic. Over the years the picnic was held at various locations in the Annapolis Valley, where he had lived much of his life and been so politically active. One year I accepted an invitation to speak to the group about my grandfather, but when I began working on my talk I was embarrassed to find out that I didn't know nearly as much about his life as I felt I should.

During the following year I went to the Archives at Dalhousie to search out the material from my grandfather's life. In a tattered cardboard file box that had been tucked away in some corner of the Archives, I found an entry visa for the Soviet Union dated 1923, letters Roscoe Fillmore had written his wife, my grandmother, from Siberia, references to the time he had spent with American socialist "Big Bill" Haywood, and more than two hundred pages of a partly completed autobiography. With this amazing wealth of material close at hand, I soon decided to write my grandfather's full biography. My research revealed a surprising amount of material by and about Roscoe. He had written more than fifty articles for two socialist newspapers soon after the turn of the century, another two hundred columns for Cape Breton's *Steelworker and Miner* over a period of nearly twenty years, four books, and countless letters to the editor. The biggest breakthrough I had was to obtain most of his RCMP security file by filing requests over a period of two years under Access to Information legislation. I was shocked to find that his file contained more than five hundred pages.

I have written this book for both personal and historical reasons. While I have tried to assess my grandfather's life critically and honestly, I am proud of his accomplishments and his lifelong commitment to social and political justice. But I also want to tell the story of the small socialist movement of the Maritime provinces, as seen through the eyes and experiences of one man who was intensely involved in and committed to the movement for more than forty years.

I feel that this chapter of Canadian history has been sadly neglected even by those purporting to explore the socialist and communist movements in Canada. Many writers working on books about the "national" socialist/communist movement have not done sufficient research concerning the movement in the Maritimes. The assumption seems to be that nothing happened in the Maritimes or, if it did, it wasn't very important. I hope that, in a small way, this book will help to dispel the myth of the Maritimes as a conservative monolith where no one had the courage to offer resistance to capitalist exploitation.

New Brunswick and Nova Scotia

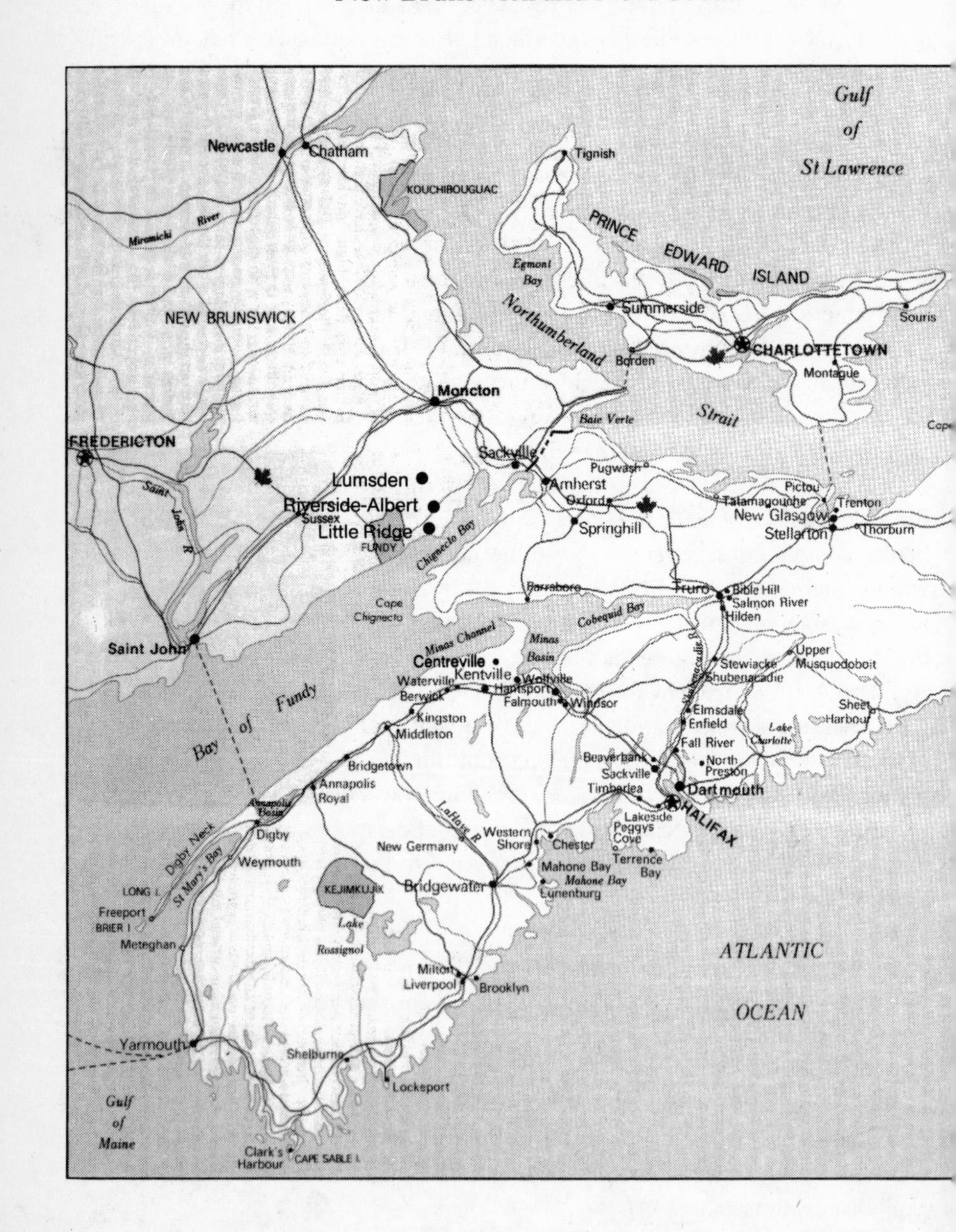

ONE

In the Backwoods of New Brunswick

TUCKED INTO THE southeast corner near the province's border with Nova Scotia, Albert County was the smallest and least populated of New Brunswick's counties. In the 1870s the county's twelve thousand residents lived in small, relatively isolated villages along the rivers and inlets that fed into the Bay of Fundy. There was no road to Moncton, the nearest town. Two steamers stopped at villages along the shore, affording settlers their only link with Moncton and Saint John. A small railway system built to serve the forest industry was not connected with outside communities. The main form of local transportation was by horseback or carriage.

Until 1849 Albert County and many other parts of New Brunswick had benefited from a preferential trade arrangement with Britain. In the best years as many as six hundred ships laden with wood had sailed across the Atlantic, but when preferential conditions were removed the timber trade had slumped badly. To make matters worse, another mainstay of the economy, the construction of wooden ships, declined dramatically in the second half of the century with the advent of iron ships.[1] By the 1870s the area's economy was in decline.

John and Elizabeth Fillmore provided for their family by operating a small mixed farm on an inferior piece of land at Little Ridge, Albert County, a tiny village about forty kilometres southwest of Moncton. To supplement their meagre income, John and his two oldest sons worked harvesting timber during the winter while Elizabeth—in addition to bearing and raising eight children—did farm work and most of the household chores.

When the New Brunswick government announced in 1876 that it would give away free one-hundred-acre plots of land to settlers willing to live in the interior of Albert County, the Fillmores seized the opportunity. The offer was part of a government plan to open up wilderness areas for lumbering. In return for the land the government was asking settlers to carry out certain land-clearing, building, and residence duties and to assist in the lumber harvest.[2]

The Fillmore's new property was located on a plateau near the top of Lumsden Mountain, a nine-hundred-foot-high ridge of forest-covered rock that lay about fifteen kilometres inland. Elizabeth's brother and his family as well as two other families were already living in the village, called Lumsden Settlement. So in the spring of 1876, John, at age fifty-four, and Elizabeth, forty-two, and seven of their eight children—the oldest daughter was to be married and stayed behind—loaded their few possessions into an ox cart and made their way through the New Brunswick forest to Lumsden Mountain. It was about ten kilometres from Albert, the nearest village, to Lumsden, on a route that took the Fillmores through thick underbrush and swamp to the base of the mountain and then up the steep mountainside to the plateau.

John and Elizabeth clearly relished the move to a more rugged environment. They were seeking a simple life of earning their living from farming the land, and they preferred to live in a small community away from the rest of society. John and his oldest sons, twenty-two-year-old Beverly and fifteen-year-old Willard, began to clear farmland from the wilderness. They cut trees, trimmed them, and used their ox to haul the logs to a small clearing where they built a log cabin. While Elizabeth went about the business of transforming the rough cabin into a home, John and the boys cleared and burnt off a small piece of land and planted patches of oats, barley, and a few vegetables among the stumps still in the ground. By 1880 eighteen families had received land grants and set up homes in the settlement, following the same backbreaking process.

Willard Fillmore was a particularly hard worker. A few years after the family arrived he managed to clear his own land and began to build his own house. At the same time he started to think about finding a wife. In 1885, at the age of twenty-four, Willard met and married twenty-one-year-old Maude Alward of Havelock, a small village about forty kilometres to the north. Two years later, on July 10, 1887, Maude gave birth to their first son, Roscoe Alfred Fillmore.

Like most other couples living in the backwoods of Albert County, Roscoe Fillmore's parents barely eked out a living. Willard looked after their small farm, planting vegetables and grains and tending two or three horses and the ox. He helped with the lumber harvest and sometimes worked as a labourer in Albert, walking the seven kilometres to work in the morning and returning home in the evening. Maude tended the vegetable garden, raised chickens, milked the cows, churned butter, did the often onerous housework, and looked after young Roscoe. When Roscoe was two years old, a sister, Ellida, was born. Two years later a second sister, Flossie, arrived.

According to his unpublished autobiography, Roscoe didn't realize as a young boy just how poor his family was. They always had food, though sometimes in short supply. Roscoe said his mother had to make moccasins for the children because they had no money to buy shoes. "In those days," he wrote, "it was a struggle, and a grim one, to raise enough cash to pay taxes and [buy] the few other items that could not be bartered for." They used their eggs, potatoes, and lumber supplies to pay for tobacco, sugar, tea, and molasses. Roscoe would be ten years old before he saw an orange or banana.[3]

Despite the poverty, the Fillmore's one-and-a-half-storey frame house was a cut above the log cabins of most families in the area. Maude had been born in a frame house and Fillmore pride would not permit a return to a log cabin, even though it would have been more economical. When Roscoe was growing up, only the lower floor was finished, and the outside was sheathed with huge slabs of birch bark held in place by strips of board. Willard always intended to shingle or put clapboard on the house but never had the money.

Lumsden was a quiet, safe place for children. When Ross—the name family members gave to Roscoe—was four years old he could walk the half-kilometre distance to his grandparents' home. The road through the middle of the settlement was straight and bordered by stone walls. Neighbours could see him toddling along, his bright blond hair shining in the sunlight. On the way was the Keillor's sugar woods, where Ross had his first taste of maple syrup. At his grandparents' place Ross watched Grandmother Elizabeth make tallow candles from scraps of fat after Grandfather John had butchered a steer, and he helped his grandmother gather high bush cranberries out back of their house so she could make jelly.

Ross's favourite relative was Grandpa John, a stout man with broad shoulders and a full white beard. He was a quiet, private person with simple wants. He loved to work in the garden. While John was a hard worker, he was not considered to be a particularly good manager of the farm. He had the appearance of being stern but Ross found him to be kind: he would hide candies in his pockets for Ross and the other kids to find. He also had a reputation for being very strong. It was said he once divided a 200-pound barrel of flour into two bags and carried them the full ten kilometres from the general store in Albert to his home in Lumsden.

In many ways Grandmother Elizabeth was the opposite of her easy-going husband. Roscoe remembered her as an impatient, stern woman with a quick temper—a view shared by other members of the family. In keeping with the style of the time she wore her hair tied tightly at the back of her head, and she had a serious look about her, with piercing eyes and drawn cheeks. Elizabeth was proud and ambitious and, like John, capable and hard working. She was a hardy, determined woman who gave birth to eight children over a span of eighteen years and raised them all to adulthood—a remarkable feat in an era that had an infant mortality rate of almost one in five.[4]

Elizabeth was a tower of strength in an emergency. Once John was thrown onto a pile of rocks by a runaway horse. He severed an artery and the blood was spurting, but Elizabeth calmly sat down beside him, found the place to apply pressure, and stopped the bleeding. She shouted for one of the boys to go to Albert for the doctor. The boy had to ride a horse for the ten kilometres, and then the doctor's buggy had to travel back over the steep and rough mountain trail. Elizabeth attended to the injury the whole time, probably saving John's life.

❧

The fundamentalist doctrine of the Baptist Church dominated the lives of most of the settlers in Lumsden—for many people in the community, the church provided the only social activity. The Baptist religion adhered to sacred covenants with God that told how members should behave, and church members who acted badly were disciplined through public censure or, for more serious offenses, exclusion from the church. Church members were supposed to watch for the errors of other members, a practice that often led to what Roscoe later viewed as puritanical interference in the lives of other people.[5]

For Grandmother Elizabeth, faith in God was clearly the most important element in life. Ross grew up to abhor his family's strict observance of what he believed was an outdated, backward religion. "Their natural bent," Roscoe wrote later, "was to 'sin'—as their little peccadilloes were seriously considered—and the harsh, uncompromising faith with its 'sins and punishment' theories, kept them in a constant guilt complex." The religion held that pride and worldly things were of no importance, and many of the followers of the church believed in the existence of a fiery hell. Apparently, the puritanism of Grandmother Elizabeth's Baptist upbringing was so strong that she even objected on religious grounds to John smoking a pipe.

"She had an ingrained conviction that anything a human might find pleasurable must be sinful," Roscoe wrote.

Many aspects of life in Lumsden were primitive. Supplies could be obtained only by barter or by purchasing them in Albert. The government refused to build roads to serve such a small, isolated community. There were no medical services. The settlers had no mail service for more than ten years after the village was established. A small school was built to be shared with a neighbouring backwoods village, Caledonia, but the parents found it difficult to get a teacher who would stay in the community. There was no police force in Lumsden, and the isolation of the community led the residents to resort to hillbilly justice to resolve their differences. On one occasion, in 1888, the Fillmores and Elizabeth's relatives in the Tingley family were involved in a clash with a family called Wilson. The feud caused a sensation and received extensive coverage in the Albert newspaper.

According to testimony given in court, the episode started when James Wilson, the family's father, bragged that he was a great fighter. The Albert newspaper, *The Maple Leaf*, reported that Wilson said, "I won't fight the crowd, but I'll fight the best of 'em from the forks of Caledonia road to Rat-tail Brook, rough and tumble, the best way I could."[6]

A few days after Wilson's boast, one of the Tingley boys got into a fight with Wilson's son Malcolm. Mrs. Wilson broke up the fight, but within minutes more trouble was brewing. Malcolm Wilson testified that the Tingleys and the Fillmores wanted to beat him up because he had struck the Tingley boy. *The Maple Leaf* reported that James Wilson stated in court that on the Thursday evening of the previous week he had heard a crowd coming toward his house. Wilson and his wife, two sons, and daughter went out to meet the crowd and order the trespassers off his premises. Wilson's wife carried a lantern, "his daughter had a stick, James had a pitchfork, Walter had an axe—and Malcolm stayed in the house." Words were exchanged and Willard Fillmore struck James Wilson "midship on the left side of the head" with a club. According to the newspaper, "The Wilson girl struck at Fillmore and he took the stick out of her hand and struck at her."

Willard and the four men with him then left the Wilsons, but Willard was later charged with assault and battery. He was convicted and ordered to pay court costs and a $20 fine—a large amount of money, more than Willard could earn in two months working in Albert.

As a youngster, Ross had his own share of misadventures. On one occasion, several children were playing in Grandpa John's backyard with an axe, cutting turnips to feed to the cows. Ross held the turnip and a little girl swung the axe. All of a sudden there was blood and screaming and Ross's index finger hung to his hand by threads of skin. Grandmother Elizabeth rushed to the scene, accompanied by a Baptist preacher, Sammy Clark, who happened to be visiting the family. It seems that Elizabeth, with her exaggerated opinion of the wisdom of the clergy, went against her own better judgement and followed the Reverend's suggestions about splinting and bandaging the wound. But Rev. Clark insisted on bandaging the finger too tightly. Within a few days the wound was infected and the whole finger had to be removed by a doctor.

When Ross was five or six, he watched his father and grandfather take part in the log drive on Crooked Creek, the river that passed near the settlement. Each fall and winter the Prescott Lumber Company worked the forests in the area, cutting and trimming trees. In the spring, when the snow melted and the run-off began, men such as Willard and John Fillmore were involved in the log drive, sending the logs downstream to sawmills. In return for a small amount of money, the men faced the unpredictable dangers of the log drive. Many a log-driver lost his footing on the slippery logs and was badly injured or even crushed to death on the rocks by the powerful force of the logs rushing downstream.

For its part, the Prescott family became wealthy from its involvement in the lumber business. Conservative G.D. Prescott was Albert County's member of the legislature, and it was his family's company that had influenced the government to give families plots of land in Lumsden so the area could be opened for lumbering. "Mr. Prescott is the wealthiest man in Albert," said a brief history of the village later published in *The Moncton Daily Times*, "and gives employment to a large number of men to whom he is ever kind and courteous. His domestics are well provided for and the hospitality of 'Maplehurst' [his home and the most picturesque spot in Albert] is known far and near."[7]

The manner in which the Prescotts became wealthy through their connections with the government was not unique in New Brunswick. For many years the province's vast forest resources had been exploited for the benefit of a few wealthy families and individuals. During the first half of the nineteenth century the large forest operators used their government connections to obtain monopoly privileges. Later the province's Crown lands were opened to reckless and wasteful timber operators who were required to pay their cronies in the government exceptionally low stumpage rates for the privilege of exploiting the forest resources. Even though it was painfully apparent that the soil beneath the forests was not fit for agriculture, few people at the time regarded lumbering as anything more than a transient industry.[8]

❧

The Fillmores and the other families in Lumsden were unable to earn enough money from the forest harvest to prop up their ailing farms, and the harsh environment of the New Brunswick wilderness soon began to take its toll. Farmers noticed that their crops were deteriorating. They had little knowledge of soils and fertility so they didn't realize that the crops had done well at first because of the ashes left in the soil after the fields were burned. When the nutrients created by the ash were used up, and with no fertilizers being used, the thin forest soil could not sustain healthy crops. The quality of the crops fell off substantially.[9]

When the crops failed, the families started relying more on raising cattle, but this meant they had to clear large areas for grazing. To supplement their supply of feed, the oxen and small horses, half-starved and weak, could haul only very small loads of hay up the side of the mountain from the Shepody marshes, some twelve kilometres away. It was a near-impossible task. Some new families came to Lumsden, but many of the young people who had been born there moved away because they found the life too demanding.

The situation became even more difficult for Willard and Maude after their second son, Frank, was born in the summer of 1893. Willard considered getting work as a labourer or farmhand in one of the villages near Lumsden, but few jobs were available. The economy of Albert County and, for that matter, the economy of most of the province were in poor shape. New Brunswick politicians believed that the province's entry into Confederation and the construction of railways across the country would allow it to benefit from increased trade. Maritime industrialists tried to bring new industries to the region, but the economic growth occurred mostly in the cities and larger towns such as Moncton and Saint John, while areas such as Albert County reaped little benefit.[10] Then, in the early 1890s, a consolidation of industry across Canada meant closings and cutbacks for many Maritime businesses. New Brunswick became a small extremity of a vast new nation and was destined to suffer further decline.

In comparison to Albert County, the economy of Maine—where the Fillmores had several relatives—was doing well. During an earlier summer, when Willard had worked for a farmer in Maine, he had earned $3 a day and board. When he worked as a labourer in Albert, he earned less than $1 a day and board. So, not surprisingly, thousands of families were leaving New Brunswick for the United States. Between 1881 and 1891 the population of Albert County had declined from 12,329 to 10,971, continuing a pattern of out-migration from the Maritimes that had begun in the 1860s. More than a quarter-million inhabitants departed the region during the last thirty years of the century, leaving a population of 893,000 in 1901.[11]

There were few barriers to migration between the United States and the Canadian provinces. Many Maritimers travelled to the States for work and, in the West, Americans sought land in the spacious Canadian Prairies. Roscoe's four aunts and uncles who had already left Lumsden for the United States seemed to have made a much better life for themselves than those who stayed behind. Two aunts were in Boston: one a schoolteacher and the other a housewife. A third aunt was married to a contractor in Portland, and one of Roscoe's uncles was establishing a career as a building contractor in the same city.

Although Willard might have found work in one of the cotton or woollen mills in Moncton, or perhaps in the iron industry in Saint John, he had no experience at working in industry. Moreover, he disliked the idea of living in a large town or city. He was more interested in something closer to his family's traditional way of life.

❧

Willard and Maude decided they had little choice but to join the flow of migration. When Willard's earlier work on the farm in Maine led to an offer of a permanent job, he accepted. In October 1893 the family left Lumsden for Gorham, a small town near Portland in western Maine.

Ross liked Maine. There were long trains to watch and, for the first time in his life, plenty of apples to eat. But only five weeks after their arrival, Maude became ill. Roscoe later wrote, "Dad worried, and finally called in a doctor who said 'If you want to get your wife home alive, you better go soon.'" Maude was diagnosed as having tuberculosis. "The roof had fallen in on our family! There was no cure for consumption in those times, at least not for poor people. It was a death sentence."

The trek back home started. "The first night on the train, from Portland to Saint John, was a nightmare," Roscoe wrote. "There wasn't money for a sleeper so we settled in a coach. My mother rapidly became very ill, delirious as the long night dragged on. Workingmen in that coach came to my father and begged to be allowed to pay for a sleeper for her but the Fillmore plus Tingley pride forbade." For Roscoe, looking back later, this was a false pride. "Such help is not charity. It is proverbial that the poor have helped each other or none could have survived."

In Saint John the Fillmores stayed for several weeks with Ross's Aunt Arminta. From the start, no one expected Ross's mother to live. Death from tuberculosis was a lingering, painful affair. Some people believed that the disease was inherited; others that it was divine retribution. Roscoe later wrote that while his family members had been conditioned by their fundamentalist upbringing to accept Maude's illness as divine retribution, they were badly confused by the fact that a person as sweet and caring as Maude would be struck down.

Ross's father had only $5 left when the family returned to Albert County, but the neighbours in Lumsden pitched in with food and support. Around Christmas, with Maude extremely ill, Grandma Elizabeth—"true to her fundamentalist faith," Roscoe wrote—insisted that the family tell Maude she was dying. "She was mortally ill, had four children ranging from eight months to less than seven years and she must be told that she was leaving them within a few days. My mother was only twenty-nine years of age. Work, childbearing, poverty, had worn her out long before her time and now she must be tortured with a brutal statement that couldn't do other than add to her sufferings." Ross never forgave their behaviour.

Maude Fillmore died on January 26, 1894. She was not buried in the Lumsden cemetery because the family feared the village was on its last legs and they would not leave her remains in a place that might be abandoned and taken over by the forest. Maude's body was taken by wagon to Albert, where she was buried in the Baptist cemetery.

Ross was six years old, and not surprisingly his mother's death had a strong impact on him. He grew to think that if the family had not been poor it could have paid for treatment in the United States to cure the illness. The same poverty that was a blight on the lives of most of the New Brunswick families that Ross knew had taken away his mother. But in later years Roscoe could hardly remember his mother. "Of course the few memories I have of her are pleasant so I am a biased witness," he wrote, "but my uncles and aunts all loved her from the first and have told me again and again that she was pretty, even handsome, gentle and loveable."

After his mother's death, Ross and his family moved into a house his grandmother rented in Albert. Unfortunately for the Fillmores, the difficult times did not end. During the early summer of 1894 Ross's grandfather came from Lumsden to Albert for a visit. Unknown to anyone, Grandpa John had contracted typhoid fever, perhaps from drinking contaminated water. After he returned home to Lumsden, he was overtaken by the fever and became seriously ill. John Fillmore lived for only a few weeks more, dying in August at the age of seventy-two. Many years later Roscoe wrote that he was sure he inherited his love of the soil and gardening from Grandpa John.

Within two weeks of Grandpa John's death, the fever struck Grandmother Elizabeth, Ross's Uncle Arthur, and a child of another closely related Fillmore family. By early September, two other family members, including Ross's aunt, who had been nursing those already sick, were ill with the virus. With family members quarantined and unable to earn a living, Ross's family was nearly destitute. "The Fillmores, poor but proud," he wrote later, "had to swallow their pride and became the objects of charity. The community raised money to keep us alive while the disease raged." To protect the young, all the Fillmore children who had not already left Lumsden were sent to live in other communities with relatives who had not been hit by the epidemic.

In those times typhoid was one of the most dreaded of all diseases. Epidemics were an annual occurrence throughout Canada, and a large city such as Montreal had thousands of cases annually. The death rate was high. The fever was caused by bacteria in contaminated food or water, and the epidemics continued to recur for years, until better sanitary conditions were introduced and water systems were treated with chlorine.

Oddly, this particular epidemic was confined almost entirely to the Fillmore family. After Grandfather John Fillmore was struck down, the others all contracted the fever while nursing the ill, mostly because of their ignorance about how the illness was transmitted. The Fillmores didn't seem to understand what to do to stop the epidemic from spreading. They had no particular medical knowledge about the treatment of the disease other than the traditional method of denying victims both fresh air and cold water. Many people in New Brunswick still clung to a more or less modified form of this treatment. The family almost certainly violated the regulations of the Public Health Act, which called for a rigid quarantine of anyone with typhoid fever for a minimum of three weeks. Another regulation was that only people treating the patient were permitted into the sick room.

For a period of several weeks the Fillmores tried in vain to obtain the services of a qualified nurse to take over the care of their sick. The nervous editor of *The Maple Leaf* newspaper wrote of the serious concern in the community that the epidemic might spread. Finally, in early September, the Fillmores obtained the services of Belle Stevens, a trained nurse who became the heroine of the family. Stevens, an intently energetic woman, worked almost around the clock to save the lives of the Fillmores. A romance developed between her and Willard, but she was unwilling to take on a ready-made family of four small children and Willard refused to break up the family, so there was no marriage.

The illness of several members of the Fillmore family had advanced so far that Belle was unable to save them. Three more died in October, including Ross's Uncle Beverly, age forty, who left a wife and three children. Among several people newly hit by the fever was Ross's father, Willard. When Beverly died, the chairman of the Board of Health enforced the law against public funerals in typhoid fever cases. *The Maple Leaf* editor wrote, "It was a sad sight on Tuesday to see the mortal remains of the late Beverly Fillmore conveyed to their last resting place, unaccompanied by mourners."[12] Beverly's family was left destitute. The family's sheep and cattle and Beverly's personal effects were sold at auction by the sheriff. The mortgage on the property was held by H.R. Emmerson, Albert's Conservative member of the provincial legislature and later a federal cabinet minister. Following the proceedings, Emmerson instructed that the sum realized be paid over to Beverly's widow.[13]

In one year Ross had lost eight members of his family, including his mother, his grandfather, an uncle, and an aunt. Both his father and his grandmother recovered from the typhoid. The experience gave Roscoe a life-long fear of disease, not for himself but for the many children and grandchildren who would later be born into the Fillmore family.

The typhoid epidemic was just about the final blow for the little settlement of Lumsden. The inhospitable backwoods mountain environment and the typhoid epidemic were more than the remaining families could take. In November 1894, with people still dying, *The Maple Leaf* reported that thirteen families—including three families of Fillmores—had moved out of the settlement. The remaining members of Ross's family moved to Albert and other villages on the Shepody River. Some families moved to Maine and Ontario.

The Lumsden post office was closed and the one family that remained in the village for the winter of 1894-95 left in the spring. The settlement was deserted, and the spruce trees, the fir, and the moose and deer of Lumsden Mountain began the long process of reclaiming what was rightfully theirs.

TWO

The Germination of a Radical

BADLY SHAKEN BY the suffering of the typhoid epidemic, the Fillmores struggled to rebuild their lives. Elizabeth Fillmore looked after the housework and three of the children—Ross, Ellida, and Frank—while Flossie went to live with relatives. Willard Fillmore travelled from village to village, picking up whatever work he could find. But Albert County provided few employment opportunities for Willard, who had only general work skills and a Grade 5 education. To make it even more difficult the economy had declined even more during the fourteen years the Fillmores had lived in the backwoods village of Lumsden.

One of the few healthy industries was the lumber trade. Each year several million feet of lumber were exported to Britain by the four lumber companies in the immediate Albert area. Although working as a logger didn't pay well and was difficult and dangerous, Willard took employment in the woods. Most families, including the Fillmores, combined lumbering and farming to earn a living.[1] When such employment was available, Willard also worked at various shipyards as a construction hand, at harvesting hay and other crops, and as a general farmhand.

Willard's many jobs meant that the family moved often. Roscoe later speculated that, had it not been for the exemplary behaviour of his father and Grandmother Elizabeth, the family's many changes of residence could have led people to call them shiftless. "Dad's hard work, Grandma's strict attendance at church, the presence of all of us in Sunday School and the fact that we always in some way paid the rent and the store bill, saved us this indignity."

By being resourceful, Willard and Grandmother Elizabeth managed to put food on the table, and no one went hungry. Roscoe thought that his father and grandmother put to shame the financiers of Wall and St. James streets by feeding the large family with $25 to $35 per month. Elizabeth also had to care for her son Arthur, who occasionally suffered from epileptic seizures.

The family's moves were disruptive for Ross, who was shifted several times in the middle of a school term. In one of the villages he had to walk nearly three kilometres to school. But Ross didn't consider this a hardship. He liked school. He was fond of his teachers and appreciated the help they gave him. Ross was a good student, and when he was nine he tied for top place in his class.

By 1898 the Fillmores were permanently established in the village of Albert, which stood on the bank of the Shepody River about five kilometres inland. Across the river was Riverside, and the two communities were home to about seven hundred people. During the summer months the river was dotted with the white sails of vessels carrying goods in and out of the villages. In the background rose the forest-clad hills, the link to Lumsden Mountain. From a high hill in the middle of town villagers could see the marshlands and, in the waters just off shore, large sailing ships and steamers gathering lumber and other products for export.

Even though Albert was isolated, unlike Lumsden it had everything that local residents needed. There were five general stores that sold everything from clothing to food stuffs. There were two tailor shops, a blacksmith, a basket factory, a carriage maker, a farm implements store, five small hotels, and its own weekly newspaper, *The Maple Leaf*.[2] It was an exciting place for a small boy. Ross liked to watch the activity on Main Street. Occasionally he was treated to the ice cream that one of the stores sold on Wednesday and Saturday evenings. Ross and his friends liked to slide on the mud banks along the river and, in summer, they found strawberries to pick on the marshes. One summer a train crashed through the Shepody River bridge and Ross watched the huge billows of steam and smoke and all the commotion of rescuing the crew.

Ross was an extremely shy boy. His insecurities may have stemmed from the loss of his mother and grandfather, the two people to whom he was the closest. He was small for his age, which may have accounted for his lack of interest in sports. He was also very serious, and his favourite activity was reading. He liked nature and often spent time in the woods, picking berries or just walking. He was fond of fishing but discovered that he didn't have the patience for it and that he lacked the timing to yank the fish out of the water.

Because his father was often away working, Ross was cared for by

Grandmother Elizabeth, and her strict views and backwoods habits frequently led to problems. One time she carefully made Ross clothes for school but, having come from Lumsden where people made do with whatever they could, she failed to take notice of what the other children were wearing. Ross knew immediately that the pants were too long and baggy. When he went to school the other children shouted, "He's wearing his grandfather's pants!" He remembered it as his most embarrassing experience as a child.

Ross fought his tormentors but, not taking naturally to brawling, almost always lost the battles. No matter how much he pleaded, his grandmother would not alter his pants. "Grandma lived all her life convinced that some sort of Divine Plan existed to try and harden the souls of people and the hardship was just to strengthen us." Perhaps the teasing and bullying was good for Ross. He soon found himself bristling at injustice and bullying, a characteristic that stuck with him for the rest of his life.

❧

On one occasion, when he was about nine, Ross's quest for reading material brought him into direct conflict with the beliefs of his grandmother. When the family lived in a rented house in Hopewell Hill, Roscoe pried his way through a window into a locked room and discovered dozens of books. Without telling anyone, he took *Robinson Crusoe* and *The Clockmaker*, Thomas Chandler Haliburton's humorous portrait of the exploits of the mythical figure Sam Slick. Grandmother Elizabeth soon discovered the books and was outraged—the Bible, the Hymnal, and the sermons of Charles Spurgeon were among the few books she would tolerate. Upon cross-examination, Ross wouldn't tell her where he had found his new reading material. Declaring that the books were "trash and lies," she threw them into the fire. The book-burning helped embitter the boy against his grandmother for the rest of his life.

At other times Roscoe enjoyed his grandmother. In the evening, as twilight came, Elizabeth would put Ellida and Frank on her knee, and while Ross sat nearby in a small chair she would sing the old hymns, "Rock of Ages," "There Were Ninety and Nine," and so on through a long repertoire. On these evenings, Roscoe thought, she was often softer and more likeable than usual. The sing-songs were usually preceded by her reading of a chapter from the Bible, and Ross learned to love the beautiful musical language of the book.

As he grew older Ross began to rebel against the strict regime of his grandmother. When he was eleven, his father went away to work in Maine for thirteen months and Grandmother Elizabeth was in complete control of the household. Perhaps the biggest conflicts came on Sunday over what Elizabeth felt was the proper behaviour for the holy day. The children weren't allowed to study their lessons or do anything else that day. They went to Sunday School, they went to church, and that was it. "A squeak out of a child on Sunday was considered a mortal sin," Roscoe recalled, "and restlessness in the hard church pews, while a preacher droned utter meaningless (to me) phrases for an hour, was unforgivable. So we often battled, I refusing to be switched and she determined to hammer righteousness into me with a birch."Even before Ross was old enough to have an interest in sex, Elizabeth's opinions on the topic left the boy puzzled. One time Ross was curious about something said by a seaman from one of the ships. The sailor had boasted that "The three most beautiful things in the world are a woman's breasts, a full-rigged ship and a field of wheat, in that order." When Ross, in all innocence, repeated this to Grandma, she angrily responded, "That man should know better than to tell filthy things like that to a child!" Ross was confused by her reaction. Were a woman's breasts filthy or should they be considered sacred and beautiful?

Grandmother Elizabeth's strict religious beliefs also led to bigotry. Her attitude was that there might be a good Catholic, but it would be unusual if there were. In Albert the Catholics lived in isolation at one end of the village while the Baptists and Methodists were at the other end. Ross was not permitted to associate with Catholic children. Elizabeth also viewed divorce as an abomination. She had been known to deliberately insult the few divorcees she had met and to hold forth on the state of sin in which she believed they were living.

In 1899, when Ross was eleven, Willard returned home from working in Maine and began to play a bigger role in the upbringing of his children. Seven-year-old Flossie, who had lived away from home during most of the time since their mother had died, returned to live with the family. That winter an outbreak of diphtheria swept through eastern New Brunswick, striking hundreds of people, and Flossie was among those stricken. She died in February 1899. Her illness and death made Willard painfully aware of the fact that the remaining three children needed someone younger and perhaps slightly more easy-going than Grandmother Elizabeth to help raise them. Elizabeth was sixty-five and no doubt weary from working hard to bring up her own and then Willard's children.

In the years since Maude's death Willard had often thought of remarrying, and even before Flossie had become ill he had started courting Maude's sister, Selina, whose husband had just died. Selina, who lived on a small farm near the village of Havelock, was left with three children, and she was pregnant. During the winter Willard travelled the fifty kilometres to Havelock by horse and sleigh to see her, and early in the new year he proposed marriage. Selina, who at forty was two years older than Willard, agreed, and they were married in Havelock in March 1899. Grandmother Elizabeth went to live with her son David in Portland. Although she later returned to Albert, she would never again have control over Ross's life. But she had made an impact on the boy that he would never forget.

When Selina came to Albert she brought three of the four children from her previous marriage. Before too long, Willard and Selina had another two children of their own. Clara was born in 1900 and Mabel in 1903, making eight children in the Fillmore household. To accommodate the large family Willard managed to pull enough money together to build a modest house on the bank of the Shepody River, beside the Albert railway station. Selina turned out to considerably different than Grandmother Elizabeth. She had a happy disposition and frequently joked with the children and played harmless little tricks on them. Ross eventually found it impossible to remember when he had not considered Selina his mother. She worked constantly from early morning until dusk. Besides caring for the children, making the meals, and cleaning the house, she tended the garden, kept some hens, churned butter, and lugged water from an outdoor well for the wash. She also carded, spun, and wove the wool for the family's clothing.

The Fillmores were a tightly knit family. Most evenings the kitchen was the main centre of activity, with a large kerosene lamp placed on the kitchen table and everyone gathered around. The children would do their homework or read. Selina would sew. Willard had a ritual of reading the day-old *Saint John Telegraph*, and when he came to something he thought everyone should know he'd announce, "Special to *The Telegraph*," and read it out loud, interrupting everyone else no matter what they were doing.

Willard, by all accounts a kind and caring father, sometimes disciplined Ross, but he was never as strict as Grandmother Elizabeth had been—although he had inherited her temper. A strong, stout man, he had thick dark eyebrows and a bushy mustache. He was a proud man who tried to instill the ethics of hard work and honesty in all his children. His hard work and his dedication to his family soon meant that he was regarded as an upstanding citizen in the village of Albert. While he was not as fanatical about religion as was Grandmother Elizabeth, Willard still devoted many hours to the church. He led prayer meetings, took up collection, and was the Sunday School superintendent. He eventually became deacon of the church, a respected position in the community but one that had a major drawback: Willard often had to go around to many of the three hundred church members and try to collect money to support the financially strapped institution. Selina regularly attended the Baptist Church and belonged to the church women's society that often sent clothes to needy people overseas, but she was not as active as Willard in the community.

Willard was a strong prohibitionist and volunteered his time to support the Sons of Temperance movement. Albert County was "dry" at the turn of the century but the Scott Act, the law that controlled liquor distribution in the province, permitted people to order a consignment of liquor from a wholesale liquor dealer in Saint John and have it shipped to the village. On one occasion just before Christmas, a magistrate decided that the method people were using to receive liquor was illegal. The authorities seized most of the following shipment, and the lumbermen, sailors, and farmers who were depending on the liquor to make their Christmas a merry one arrived at the railway station and discovered that their goods were not to be had.

The liquor had been hauled to the Fillmore house, located next to the train station, and Willard, the Scott Act inspector, two constables, and several of the leading citizens of Albert had gathered there to guard it. A group of angry men who believed the liquor was rightfully theirs began to mill around the house. There was a lot of noise, but no real harm was done. Most of the men knew Willard personally, and Ross thought they respected his father's views because they knew that he really was a "dry" and not a hypocrite who voted dry and then drank secretly. Ross thought that if the liquor had been taken to the home of one of the constables, the men would have broken in and taken it.

Not all of Willard's views were traditional. When the family was still living in Lumsden, Willard took a step that was in those days almost unheard of for a Fillmore—he voted Liberal. Among most of the backwoods people and even in the villages and towns, Liberalism stood for something radical. The Conservatives traditionally governed New Brunswick, at one time for a stretch of seventeen years. Among landowners, no matter how poor, there was a fear of what the Liberals might do. Grandpa John and others in the family had fought Willard over his determination to break with tradition and support new ideas and policies.

During winter, when Willard worked at the Prescott family's lumber camps in the woods around Albert, he allowed Ross to go back to the camp after school on Friday and stay over for the Saturday work day. The boy's visits to the camp gave him a look at what it was like to work hard for a living. Ross especially liked the camp food. The camp shacks were long, low buildings, tight and warm. Ross didn't sleep very well, however. Twenty to thirty-five men slept in one room in double bunks and the air was pretty well exhausted by morning. Unpleasant smells came from sweaty clothes steaming around the stove and from feet that had gone unwashed for some time. Perhaps worst, Ross thought, was the loud snoring of the tired men. Nevertheless, it was a great adventure for a small boy, and Ross would have gone there every Friday night if he'd been permitted.

In summer, when Willard worked at the shipyards, Ross, thirteen, sometimes went along to watch the boats being built. Unfortunately, the pay at the shipyards wasn't good, and there had been a dramatic decline in the industry. Once almost every creek and river along the Bay of Fundy had a shipyard. But the shipbuilding industry in New Brunswick had peaked years before Roscoe was born.[3] When iron took the place of timber in the building of vessels, construction of wooden ships fell flat.

Ross liked the romance associated with the ships and the sea, and he loved to watch the various stages of construction as a new ship was put together. Two of the county's largest shipbuilding wharves had been at Alma on the Petitcodiac River and at Harvey, near the mouth of the Shepody River. Ross occasionally got to visit the shipyards at Harvey where, from 1875 to 1892, twenty large vessels had been built. He witnessed one of the last launchings of a large schooner there when he was about thirteen. It was a three-masted vessel, christened the *Ethel B. Summer*. Ross walked three miles from his home in Albert to join dozens of other youngsters who were allowed to go on board the ship when it was launched. When he became old enough, Ross worked with his step-brothers on the wharves, tending the conveyors that carried lumber onto schooners.

Willard eventually got full-time work on a large farm owned by the McClelans, one of the most successful and wealthy families in the county. Willard was paid a dollar a day, with certain fringe benefits. When there was a slaughter of beef cattle he could take the unsaleable livers home for his own family—a sign of the paternalism that existed in the community, which Roscoe would grow to detest. In addition to the cattle farm, the McClelans owned a lumber company, a saw mill, a grist mill, a shipbuilding firm, and a general store. They were also deeply involved in politics. During the years the family amassed its wealth, Abner McClelan was the member of the New Brunswick legislature for Albert County, and after the province joined Confederation he became one of the first men appointed to the Canadian Senate. Later he was Lieutenant-Governor of New Brunswick. At the end of his term, according to a 1910 report in *The Daily Times* of Moncton, he retired to "his magnificent home in Riverside, where he lives the life of a country gentleman."[4]

❧

One of the people who took a special interest in Ross during his early teenage years was J.H. Rhodes, an elderly Englishman who had taught music in the village and was editor of *The Maple Leaf*. Rhodes lent books to Ross and helped him understand their meaning. Ross became fond of the old man and later gave him credit for expanding a young boy's curiosity and instilling him with a social conscience. Sadly, Rhodes later became an alcoholic, lost his job, and was evicted from his rented house. His possessions were seized and sold at auction to pay his debts. Before his disgrace Rhodes had introduced Ross to the novels of Charles Dickens. "I bled inside for his heroes and heroines," Roscoe later wrote. "Dickens' portraiture of the English people under the oppression of aristocracy and mill owner, conniving officialdom and hypocritical churchman and cruel kings ground into my very soul man's inhumanity to man." Rhodes also gave Ross a copy of Thomas Paine's *The Age of Reason*, which pointed out false aspects of religion and discussed inconsistencies in the Bible. Roscoe admired Paine for the role he played as a driving force behind American independence and considered him a hero for attacking the wealthy bankers and merchants who were withholding support from Washington's army during the war of independence.

Ross also got an early education in patriotism, for Albert was an extremely patriotic village. When the Boer War was raging in South Africa, most villagers supported the British. Ross and his schoolmates loudly sang "Rule Britannia," "The Soldiers of the Queen," "Goodbye Dolly Gray," and "God Save the Queen." But after he had left school, Ross underwent an agonizing reappraisal of his sense of values over the war. He deeply regretted that he had been critical of one of the local preachers, Rev. John J. King, who doubted the righteousness of the conflict. Though Ross had not heard of socialism at that time, he later learned that Rev. King was a Christian Socialist and a pacifist. King condemned the invasion of the Boer Republic and charged that Britain had an eye on the great gold and diamond mines. King's position "was treasonable," Roscoe later wrote. "I was too young to take part in the bitter vandalism that broke out in that law abiding village, I am thankful to say, but I had the will. The vandals were from the 'best families' of the community. A pig and a flock of poultry were turned loose in the parsonage parlour, windows were broken, and clothing destroyed on the washline. The vandalism went on for months."

Around 1902 Willard started to work at the Albert Nursery, which was in the business of growing and exporting young apple trees to farms throughout the Maritimes and Quebec. The nursery sold more than twenty thousand young apple trees a year, as well as a few thousand plum and pear trees.

Part of Willard's job involved travelling by horse and buggy throughout the county, selling and delivering fruit trees to farmers. After Ross turned fourteen he often accompanied his father. They'd stay overnight with friends who would put them up and feed them. Ross began to work more often at the nursery, even to the point of missing some school. He loved the work, and the income helped the family. During summer holidays he hoed and weeded and learned to prune and stake and straighten trees. He delivered nursery stock with an old nearly blind horse. On Saturdays and after school during the winter Ross made root grafts for the thousands of apple trees sold by the nursery. He perfected the technique of grafting at a young age and in later years horticulturalists said few people could match his skills with a pen knife.

When Ross was about fifteen he began to take an interest in one of the girls at his school. After watching her for some time he decided to do something about it, but he was far more shy than most boys and couldn't bring himself to speak to her. He knew she liked him, but also knew that her parents wouldn't approve. "My family were poor, struggling wage workers while her father was a well-to-do farmer." When the girl began to sing in the Methodist choir, Roscoe started attending that church. On Valentine's Day Roscoe was determined to do something big. "I bought a Valentine that cost thirty cents, a gaudy and beautiful thing, and sent it to my girl," Ross wrote later. "I don't remember how I scratched up so much money as this in those days but I did it. I don't recall if it was ever mentioned between us or whether she ever said thank you or not but I am sure she thanked me with her eyes many times." When Ross moved to Portland a year later he wrote letters to her and, with the distance between them, wasn't so shy. In one letter he declared his love for her. This ended the correspondence, because the girl's parents thought she was too young to receive such declarations.

The Fillmore family's limited financial resources made it impossible for Ross to pursue an education beyond high school. Like most of the boys in his class he was expected to leave school and work to help support the family. Ross regretted leaving school, but he knew he had no other option but to get a job. He believed there was little opportunity in the village of Albert; he thought the village was dominated by old people whose ideas were old fashioned. So in April 1904, at the age of sixteen, Ross dropped out of grade 10 at the Albert Grammar School and went to Portland, Maine, in search of a job.

❧

As the train slowly made its way into Portland's Central Station, Ross could see the bustling city. He saw automobiles competing for space with horse-drawn carriages and men dressed in dark suits and bowler hats stepping quickly along. Commotion was everywhere. Portland was a thriving city of more than fifty thousand people, and even though it had suffered from some deindustrialization during the late nineteenth century, the town still offered many job opportunities.[5] Grandmother Elizabeth met Ross at the train station, and it wasn't long before he was settled in with her and his Uncle David, a building contractor.

Even though he had no special skills, Ross had no difficulty finding work. But he hadn't held his first job long before he got into trouble. By his own admission, Ross had become a bit belligerent, with the quick temper of his grandmother and father. When he told the patriotic Yankee who was his boss that the Canadian system of government was better and more democratic than the U.S. system, Roscoe was fired on the spot. "Get to hell back among your herring chokers!" the man shouted at Ross.

Unconcerned about his abrupt dismissal, Ross got a job as a stove-pipe fitter and then worked for awhile with his uncle's construction business. Later he found a better-paying job in the locomotive repair shop of the Maine Central Railroad as a machinist's helper and electric crane operator.

On his days off Ross liked the wide, busy streets of Portland, especially the activity on Congress Street. He saw the miracle medicine men who sold "one minute cures" for everything from rheumatism to croup to bruises, and he heard all kinds of preachers talking about the evils of alcohol and mysterious gambling dens. As something of a bookworm he enjoyed going to the large bookstores with their mountains of books and racks of magazines. Or he would go to the Bijou Theatre to see animated cartoons. Vaudeville was in its heyday and Ross never tired of watching the wonderful stunts, the chorus girls, the tumblers, acrobats, and hypnotists.

Ross discovered that his grandmother's strict views hadn't changed much since she'd left Albert. When Ross told her that he was fond of the theatre and that he had seen George Bernard Shaw's play, *Man and Superman*, Elizabeth said that whips of scorpions wouldn't have induced her to set foot in a theatre, which she considered to be a den of iniquity. "I argued and defended the idea of usefulness of recreation and entertainment," Roscoe wrote later, "but Grandma laid down the law that I would be much better employed reading *The Bible* and nothing could shake her conviction. She believed all those who had anything to do with the theatre were miserable sinners and very wicked people."

Ross was amazed at the number of licensed taverns in Portland. Remembering the temperance orators who had visited Albert and made claims about the strict controls on liquor in Maine, he was greatly surprised that in reality there wasn't even a pretence of enforcing the law. Curious and perhaps eager to rebel against his strict grandmother, Ross began to spend time in the bars, to the extent that not long after arriving in Portland he thought he was well on his way to becoming a bar fly. But a discovery made one Sunday afternoon while strolling along Congress Street was to change the course of his life.

A small crowd had gathered in Congress Square to listen to a soap-box orator. At first Ross thought the man was one of the many fundamentalists who made their living preaching the gospel, but he soon realized that this man was different. The speaker wasn't talking about God but about justice for working people. "Why is it that those who do all the hardest work in the world get the least?" he was asking.

Thinking of his own family struggle for survival in the backwoods of New Brunswick, Ross knew exactly what the man was talking about, and he became increasingly absorbed in the speech. Not all of the ideas were new to the boy. He had read Dickens, and in his own village he had seen the same contrast of extreme wealth and poverty. But Ross hadn't been able to figure out why such extremes existed. This man seemed to know why, and he managed to tie many ideas into a philosophy that made sense.

After the speech, Ross followed the man and a small group of people to a hall, which turned out to be the Socialist Party Headquarters. The man was Rev. George Littlefield, a socialist from Massachusetts. Ross collected some socialist literature and began to learn more about this different concept of how society should be organized. The Socialist Party of America had been in operation only since 1901, but Ross could see that a lot of people were enthusiastic, and he soon began to believe in the potential of socialism. He joined the Socialist Party of America and began volunteering his time to support its work.

Ross soon found that a committed socialist was expected not only to believe in socialism but also to preach it, spread its doctrines, and recruit new members. Many of the socialists, it seemed, were possessed with a fervour like that of the early Christians. Ross, shy at first, soon found himself approaching strangers and offering them pamphlets. Most Sundays he and other socialists were on the streets, at parks, or at the railway station, distributing literature. They were subjected to the sneers and contempt of people who didn't like the socialist message but, if an argument ensued, Ross could hold his own.

Surprisingly, at first Grandmother Elizabeth showed a certain interest in her grandson's new pastime. She seemed impressed by Ross's arguments about the need for change. But one day Ross made a skeptical remark about religion and Elizabeth immediately sprang to the conclusion that his socialism had made him antireligious. "From that day on she was belligerently antisocialist. A barrier had been erected and it never slipped. Of course, as relatives often will, she blamed my mistaken ideas on many things, the books I had read, bad company, absence from church and so forth."

After living in Portland for about eighteen months, Ross learned from his doctor that he had what appeared to be the early stages of tuberculosis. Elizabeth feared for Ross's life. She was convinced there was no cure for tuberculosis: This was the same disease that had killed the boy's mother. The doctor told Ross he should get more rest and work in a more healthy, outdoor environment. So Ross decided to return home to Albert. In any case by that time the initial impact of Portland had worn off and Ross had found he wasn't particularly happy there. He decided that he didn't care very much for living in a city.

❧

Ross arrived back in Albert in the fall of 1905. While he had been away the owner of the Albert Nursery, where Willard had worked for three years, had died. The nursery had reduced its volume of business considerably from earlier days, and when it appeared that the new owners might close down, Willard worked out an arrangement to buy the business. The deal included land, a couple of buildings, some nursery stock, and the company's excellent reputation. Willard didn't have enough money to buy all of the nursery's land so he rented nearby fields to grow nursery stock. The new company, Fillmore Bros., included Willard's brother Arthur, who was subject to epileptic seizures and unable to do much strenuous work.

The purchase of the nursery was a proud occasion for the Fillmore family. Owning a business was a big step up from being poor farmers, and it was possible only because of Willard's hard work and the reliable reputation he had gained. The nursery was one of Albert's most respected companies. It had been established before 1860 by Agreen Tingley, whose gardening skills were legendary. Tingley, known as the Johnny Appleseed of the Maritimes, had grown thousands of trees that were sold throughout the region as well as in Quebec and the eastern United States. The nursery sold more than twenty thousand young apple trees a year, as well as a few thousand plum and pear trees, and its stock usually contained more than two hundred thousand small trees.

Ross was planning to go to work at the nursery, but everyone knew he was not supposed to work hard because of his illness. Concerned about the boy's health, and cherishing the idea that Ross still might have the opportunity to get the education that his father never had, Willard convinced Ross to return to school and work at the nursery in his spare time.

In the late fall of 1905 Ross entered Grade 11 at Riverside Consolidated School, a handsome, huge, new brick regional school that was the pride of the entire county. He took the opportunity to increase his reading and delve into the work of philosophers, historians, and evolutionists—material considerably beyond the interests of most high school students. He devoured book after book, including the work of poet and philosopher Ralph Waldo Emerson, civil libertarian Oliver Wendell Holmes, abolitionist William Lloyd Garrison, author and reformer John Ruskin, philosopher Herbert Spencer, evolutionist A.R. Wallace, and political and social reformer John Stuart Mill.

Ross was also interested in learning more about socialism. While most people in Albert had barely heard of the new political philosophy, school principal George Trueman had studied in Germany, where there was an active socialist movement. Trueman believed that all subjects, including those not popular in the local community, should be discussed with students. So Ross was able to read the work of socialist writers, including Jack London, whom Roscoe knew for his *Call of the Wild*; Upton Sinclair, who had just published his protest novel, *The Jungle*; William Morris, author of *Socialism: Its Growth and Outcome*; and poet Percy Bysshe Shelley, whose political philosophy inspired radical thinkers.

In class Ross—who at eighteen was not the oldest student—displayed the arrogance and courage of youth, always making sure he got his say. Later he remarked that George Trueman welcomed "the verbal bombshells" that his student often tossed into class. Ross missed only five days of school and got good marks. In the winter he stood fourth in his class.

In the spring of 1906 Ross wrote a lengthy essay for Trueman, entitled simply "Socialism." The paper reflected Ross's newly found idealism. "We are fighting for a time when there shall be work and leisure for all," he wrote, "when the rich shall no longer live in luxury on the fruits of the labour of the poor; when all the good which is in the individual shall be sought after and encouraged to result in the benefit of all; when no man, woman or child shall go hungry, or live in hovels, whose very existence contaminates society."[6]

Socialism was then making its presence felt in many countries and Ross's essay was full of optimism for the future of the new political movement. He described the progress socialists were making in Austria, Italy, Japan, and Great Britain. He was particularly optimistic about the future of socialism in the United States, where the socialist presidential ticket of Eugene Debs and Ben Hanford had received almost a half-million votes in the 1904 election and socialists had been elected in several cities.

Ross was closely watching developments in Russia, where the Czarist government had sacrificed the lives of thousands of poorly trained and badly equipped soldiers in a needless war against Japan. Ross knew from reading socialist newspapers that the Russian workers were organizing, and his essay predicted: "I think I am safe in saying that within ten years Russia will have laid the foundation for a better system of government than we have at the present day." He also wrote about the relationship between socialism and horticulture, agreeing with the teachings of Luther Burbank, who argued that environment had a lot to do with the behaviour of people. Ross wrote that human nature could be changed, just as the nurseryman straightens trees when they are young so they will grow strong and powerful.

At the time Ross was interested in pursuing horticulture as a career and thought that he would eventually join his father's business. Even though he liked school and was doing well, Ross dropped out in the spring to help with the nursery and supplement the family income. But Ross found that he wasn't ready to settle into the family business; he wanted to see something of the rest of Canada.

THREE

The Great Harvest Excursion

WHEN ROSCOE FILLMORE was growing up in Albert County, harvest excursions to western Canada were an annual event. The wheat economy was critical to the nation's development, and each year a country-wide appeal was made for workers to travel to the prairies to harvest and thresh the enormous grain crops. Seeing ads for the excursions in the New Brunswick newspapers, Roscoe thought such a trip would provide the opportunity of not only earning some money but also seeing if socialism was getting anywhere in the West.

In early August 1906, nineteen-year-old Roscoe Fillmore and four thousand other workers from the Maritimes and Newfoundland arrived in Saint John to board a series of special CPR trains headed for Winnipeg. For $12 each they were able to buy a second-class ticket that would get them to Winnipeg. If they wanted to go on to Saskatchewan or Alberta they could buy an additional ticket in Winnipeg. Every sort of person imaginable jammed onto the train—farmers, miners, lumberjacks, fishermen, clerks from small towns, and schoolboys like Roscoe. Most of the men were between the ages of sixteen and thirty. There were also a few women, most of them hoping to get work cooking on the harvest farms.

The would-be harvesters travelled in the colonist coaches that had been built to carry thousands of immigrants to the West. Each coach held fifty-six passengers seated facing each other in groups of four on hard, poorly upholstered wooden seats. At night the seats became a hard bed for two while the other two passengers shared a berth above. The coaches had poor suspension and, with the constant noise on the train from arguments and merrymaking, the ever-present smell of sweat and over-ripe socks, and the uncomfortable beds, Roscoe seldom got much sleep.

Harvest specials had low priority, so they were often put on a side track for lengthy periods while regular trains went by. When the train stopped, the excursionists—mostly young men full of beer and energy—often found some way to get into trouble. Lack of food led to some of the more serious problems. No meals were served on the train, though each car had a stove for boiling water or preparing tea or coffee. Each person was supposed to bring enough food to last through the long journey, but Roscoe and many others soon ran out. So when the train stopped, many of the men would get off and loot anything they could find—small restaurants at trackside, stores in little villages, apple crops and blueberry bushes. They would rip entire bushes from the ground and take them on board the train.

Prohibition was in effect, although officials and passengers alike paid little attention to the laws during the trip west. One account of the same trip Roscoe took describes how a couple of carloads of Nova Scotians, many of them miners, brought a generous stock of liquor along but nevertheless ran out just after the train left Fort William, Ontario. When the train stopped at Ignace, Ontario, to pick up a new crew and locomotive, the Nova Scotia group charged into a saloon and restaurant maintained by the railway and took as much food and liquor as they could find. When they got back on the train they tore the water coolers off their mountings, emptied them, and refilled them with draught beer. The railway staff, badly outnumbered, didn't try to intervene. When the train arrived in Winnipeg it was greeted by a small army of constables and a dozen horse-drawn Black Marias. Some of the drunk Nova Scotians were arrested and put in the Black Marias, but their friends used knives to cut the horses free and most of the men escaped. Very few of them ended up in court.[1]

In Winnipeg Roscoe bought a ticket to Saskatchewan. When the train finally pulled into the Regina station, in the middle of the night, Roscoe and the other harvesters were dumped onto the platform. They had no idea of where to go or where to sleep, and most of them bedded down with blankets in the train station until morning. Roscoe, after six days on a train with little food or sleep, was exhausted. He was also nearly penniless and, to complicate matters, the wheat was ripening a little later than expected, so most farmers were not yet hiring. Despite his weakened condition and the fact that he was not supposed to do strenuous work, Roscoe found a job at a construction site, wheeling mixed concrete. The work was hard, but after a few days with regular meals Roscoe's health actually improved.

The system of getting hired on at a farm was haphazard. There was no central hiring office, and workers couldn't be sure that employers would live up to their promise of providing adequate food, accommodation, and pay. But one evening a farmer drove into Regina and offered $40 a month to anyone who wanted to go work on his farm, and Roscoe accepted. The man was Bob Trask, an Ontario native who owned a half section—320 acres—about twelve miles south of Regina.

When they arrived at Trask's farm, Roscoe saw that the buildings were in very poor shape. The house was small and rough, and the barn was only a series of poles over which the grain thresher had blown a mountain of straw. Roscoe soon discovered that the farm's only water hole, over near the barn, contained hundreds of tadpoles. Whenever Roscoe went to the deep hole with the buckets, he would momentarily drive the tadpoles away with a splash and then dip a bucket down. A few weeks after his arrival, a couple of other harvesters cleaned out the water hole and found several dead gophers in various stages of decomposition.

The food at the Trask homestead was not up to the standard that Roscoe had hoped for. The principal diet was bakers' bread; "skilly," as porridge was called; rice; canned milk; and lots of tea. Sometimes there was prairie chicken, the only meat Roscoe had during the time he was there. Roscoe also couldn't quite get used to the lack of trees and the flatness. On clear days he would look out from the Trask homestead and all he could see, faintly in the distance, were poplar groves—"bluffs," as they were called—near Qu'Appelle, about thirty miles to the north. He wrote, "In those days the prairie was a lonesome looking country and by the time dark came I was lonesome, homesick, tired and sick of drinking hot or lukewarm tea and craving a drink of something real cold."

Nevertheless, Roscoe took a liking to Trask, who lived alone in the small frame house. There were also some aspects of the prairies he liked. At night when he lay outside looking at the sky, the moon had never looked so big. As the grain crops ripened they were a sight to see. For hundreds of miles the fields were as level as a floor.

The harvest was done in two stages—stooking and threshing. A couple of hired hands cut the long, golden shoots of grain, which were then tied into bundles, or sheaves, by a binding machine drawn by a team of four horses. After that the stooking began—the stacking of the sheaves with the grain ends up to keep them dry and off the ground. Then the grain was ready to be threshed—the process of separating the tiny kernels of grain from their stocks. At this stage a huge threshing machine run by steam and resembling an early locomotive arrived at the Trask farm.

A crew of twenty-eight men or so was needed to run a thresher. Hundreds of the giant, strange-looking threshing machines roamed the prairies. A small steam locomotive with broad flat wheels to help it cruise across the land hauled a water tank and a caboose where the men lived. Following behind were eight teams of horses pulling the racked wagons used to haul the grain to the thresher. When this huge ensemble approached a village the whole population turned out to watch. According to Roscoe, the villagers would tag along after the curious machine and horses, creating a kind of "triumphant parade" from village to village. "In many places bridges had to be strengthened as the country roads had never before seen such heavy vehicles."

Roscoe's job was to drive a racked wagon that was pulled by two horses and hauled the grain to the thresher. His work was dependent on "field pitchers," who used pitchforks to throw the sheaves of grain up onto the wagon, where Roscoe stacked the load. On many days there were not enough field pitchers to go around, so Roscoe and the other drivers had to get down off their rigs, pitch the sheaves up, and then get back on the wagon and stack the load properly. It was an exhausting, nearly impossible task.

Roscoe and some of the other drivers soon realized they had to do something to get the crew foreman to hire more field pitchers. They decided that when they went to the fields one morning they'd refuse to do the work of the pitchers. That day, since no pitchers showed up, no grain was being loaded. But when the boss jumped up into his buggy and started to shout at them, the other drivers jumped off their wagons and started to load. Roscoe was disgusted and determined to stay on the rack, waiting for pitchers. He wrote, "The boss must have been a rather reasonable person for instead of firing me on the spot he demanded the reason and was told, whereupon he promised to go to town immediately for more field pitchers and he did so. In the caboose that night several of the drivers who had promised to hold out cursed me for 'almost losing everybody's job.' I have seen similar affairs many times since. Workers are still groping for an understanding of solidarity."

Roscoe was disappointed that only a few of the workers he met believed in unions or knew anything about socialism. He would have liked to have belonged to a union, but agricultural workers were difficult to organize. The jobs tended to be temporary, so the men would tolerate the poor working and living conditions. Some transient labourers in Western Canada belonged to the Industrial Workers of the World, but the IWW's Agricultural Workers Organization never did make the same inroads in Canada that it made in the western United States, where it had eighteen thousand members in 1916.[2]

During the threshing period the men frequently worked sixteen-hour days. There were serious injuries and deaths from limbs being caught in the farm machinery and from boiler explosions. Conditions in the fields were often hot and unpleasant. The men had to cope with swarms of flying ants, which invaded the fields on the hottest days in late August and September. The ants crawled into their eyes, ears, noses, and under their clothing—and they could bite. If they weren't driven from the fields by these annoying creatures, the men would sometimes crawl into the stooks and straw-stacks or anything else that offered protection.

The outfit that came to the Trask farm threshed grain throughout the district, and Roscoe started to travel with them and see what kind of life the immigrant settlers had carved out for themselves on the prairies. Once the outfit visited a Hungarian village of about a dozen families. The Hungarians were noticeably upset about the lies of Canadian Immigration officials, steamship officials, and the railway companies. "There had only been one thread of truth in the whole story—free land," Roscoe recalled. "The free land was here but the people who flocked in watered it with their blood, sweat and tears. The moment they hit the prairie they became the prey of railways, implement dealers, lumber dealers and the hangers on....

"It was made practically impossible for an individual farmer or group to ship their own grain—they must sell to the elevator companies and there was no competition between these outfits, all offered the same low prices. The authorities must have known what went on but there was never a peep from them.... The blatant lies wrapped up in the official immigration policy and practice is a blot on the history of Canada."

By late October most of the threshing at the Trask farm was completed. By then the nights were getting cold and there was a bit of snow. Roscoe and the other harvesters from the East had to decide what to do next. Surprisingly, about 30 per cent of the 1906 harvesters stayed on the prairies and set up their own homesteads.[3] But other than homesteading there were few jobs available on the prairies, so most of the workers either went on to British Columbia, where they planned to work in the forests, or returned home.

Roscoe decided to head home. His pay probably totalled more than $100—a lot of money back in Albert. He got Bob Trask to sign his railway ticket—the return trip back to Saint John only cost $18 if a special coupon attached to the ticket was signed by a prairie farmer, certifying that the person had worked at least thirty days in the harvest field.

Roscoe arrived back in Albert during the last week of October. His half-sisters, Mabel and Clara, recalled that, to the horror of their mother, Roscoe had to have all his clothes washed carefully because he had picked up bed bugs and other tiny creatures during his travels. Very little had changed in Albert and, after a few weeks, Roscoe became restless. There was no work for him so he went to Portland, where there were relatives to stay with and where he got a job as a carpenter. But he soon decided that he didn't enjoy what he was doing. He hoped that, with his horticultural experience in Albert, he could get better work in the rapidly expanding United States nursery business.

❧

Roscoe learned through family connections that Brown Bros., one of the largest nurseries in the United States, was hiring workers at its Rochester, New York, location. Rochester was the centre of the country's nursery business, and Brown Bros. was an excellent place to learn the latest methods of the trade. Roscoe went there in Spring 1907 and got a job on the checker's desk, where he helped keep track of the shipments that went out to the towns and villages along the eastern seaboard. Later he worked with crews planting imported stock in rows in the nursery's fields. Stock from Holland, Belgium, France, Germany, and Japan was shipped to the company, and Roscoe questioned the knowledgeable Dutch, German, and Swiss nurserymen all about the new, strange plants.

Soon he was acquainted with hundreds of species of plants that were either unknown or less developed in New Brunswick. One group of plants in particular fascinated him: a collection with broad, dark green, leathery leaves that came out of the cases shipped from Europe. These plants slightly resembled the stiff brushy shrubs that as a child Roscoe had seen covering the pastures of Albert County each spring. When the plants blossomed in late May the fields were covered with the rose-purple flowers. Roscoe learned that the plants were a species of azalea, a close relative of the rhododendron, and that they grew all over Canada and New England. The rhododendrons arriving from Europe were distant cousins of the plants in Albert County, and when they blossomed they revealed glorious colours and a profusion of flowers. Roscoe's infatuation with rhododendrons and azaleas lasted a lifetime.

Another popular plant of the time was the baby rambler rose, with its huge clusters of flowers. Imported from Holland, they became a craze. Public parks and private gardens planted them as fast as they were available. Roscoe sent rooted cuttings of the baby ramblers to his father at Albert Nurseries. Probably the first baby rambler roses ever grown in the Maritimes, the new roses proved to be hardy and were sent out to other communities.

Rochester was a city noted for its great parks, and Roscoe spent many Sundays in them, learning more about unfamiliar plants. He had not known that there were so many varieties and colours of lilacs. On the slopes of Irondequoit Bay he saw his first vineyards. In the parks he saw rhododendrons, azaleas, and Japanese double-flowering cherries. In later years he would develop some of these exotic plants for the colder climate of the Maritimes.

In the summer of 1907, after the spring shipping had been completed, Roscoe was laid off from Brown Bros. Instead of returning to Albert, Roscoe and a friend from the nursery decided to go on the harvest excursion. They bought tickets at the CPR office in Rochester and took a boat across Lake Ontario to Toronto, and caught a train west. Later, after Roscoe's second season of work as a crop harvester, he headed further west to British Columbia.

❧

By the early winter of 1907, when Roscoe and several other crop harvesters arrived in Field, British Columbia, a depression had crippled many industries and set thousands of transient workers adrift all over Canada and the United States. Still, the men believed they'd find work on the construction of a railway tunnel through the Rockies. They'd heard that the CPR was rebuilding a part of the railway line in the Kicking Horse Canyon because the line was too steep to operate efficiently or safely.[4]

Most of them still had a few dollars left from the harvest season, so they went to the Strand Hotel and got rooms. The next morning Roscoe and several other men walked three miles up the mountain to the construction camp. As Roscoe could see as he climbed to the campsite, the new line was an amazing construction feat. The grade of the original tracks was so steep that four engines were required to haul a short freight train up the mountain. To reduce the grade by half the CPR had to double the length of the line, which was possible only by constructing two massive spiral tunnels in the mountains.

Roscoe worked for the first couple of days cleaning away debris from the approaches to one of the tunnels. Then he was sent into the tunnel to work as a mucker—a particularly dirty and dangerous job. He had to pick up rock and debris that had been dynamited at the face of the tunnel and load the material onto horse-drawn hand carts that hauled it away. While Roscoe and the others cleared away the debris, men who had experience with explosives drilled and loaded holes in the rock-face with explosives. Between shifts the explosives were set off. Then the muckers and the drillers came in again and the whole process was repeated.

The construction bosses had a reputation for pushing the workers as hard as they could. Roscoe, who thought he was tough and used to hard work, almost didn't make it through his first shift. Later he wrote: "In the middle of my night shift I had to sit down. My hands felt cramped on the shovel, my back was almost useless and I was dizzy. Along came the boss shouting 'Pick it up boys, pick it up!' and I could scarcely pick myself up. He hesitated a moment, then shouted at me 'What's wrong, you hurt?' I stumbled over to him and explained. His bark was far worse than his bite for he said 'Well, keep moving a little 'till the end of the shift. A good sleep will make you okay.' He saved my bacon. I could easily have been thrown out on the grade, sick, exhausted and unable to walk to a settlement. Many men were used in that way—they were cheap and expendable."

In addition to the hard work, the men had to cope with the threat of serious illness. Each winter more than forty feet of snow fell on the mountain, melting and soaking the hard-working, sweating men through to their skin. Because of the difficult working conditions, some men didn't last long on the job. There was a saying in camp that there were always three crews—one on the job, one on its way in from town, and another on the way out of camp.

It wasn't long before the contractor that Roscoe worked for realized that with the economy in decline and men practically begging for work, the company could get away with paying lower wages. When Roscoe arrived it was paying $2.50 for a ten-hour day, six days a week. Within less than two weeks the pay was cut to $1.50 a day. The men grumbled, but there was no serious resistance. Roscoe later noted that the camp badly needed a labour organization. Workers "were cheap as dirt" and were treated worse than the horses, which were at least brought in off the grade and stabled in the worst weather. "But neither rain, sleet, snow or hail was considered injurious to men." Roscoe saw the bosses "march through the bunkhouses and haze out those who had remained in, due to the almost impossible weather. The language used was utterly unprintable and one can only marvel at the patience of workers and their lack of spirit—or many of those bosses would have left camp in long, narrow boxes."

For the first time during his two trips to Western Canada, Roscoe met socialists and labour organizers who would work for months at a time with little or no pay to organize miners. He knew men who travelled from place to place by freight train in boxcars or in the caboose to try to organize workers. It was difficult for unions to organize unskilled workers, such as railway construction crews, because the men could be fired for no reason at all and easily replaced. Although they had little success, the most spectacular campaigns were carried out by the Industrial Workers of the World. The men, mostly recent immigrants, were often cheated on their pay, and their only recourse was a trip of many miles at their own expense to complain to a magistrate, who would likely do nothing.

Companies such as the CPR were brutal in their opposition to unions.[5] There were numerous incidents of camp workers resorting to collective action to remedy specific grievances. In a CPR camp not far from where Roscoe worked, a group of forty Italian navvies went on strike over unsatisfactory conditions, threatening to use violence unless the company complied with their demands. The company made the necessary concessions, but there were few such victories for the workers.[6] There was no forced labour in Roscoe's camp but, farther north, he heard that armed guards were used to prevent workers from leaving the camps. He knew that slave-like conditions existed in some of the construction camps in the mountains, and that labour organizers died in the attempt to organize such camps.

With no union activity in Roscoe's camp his net wage for a sixty-hour week was $4.50, after the company deducted seventy-five cents a day for housing and food. Despite the low pay, Roscoe felt lucky that he was at least eating and had a place to sleep. That winter he heard stories from all over the West of men starving to death.

The construction camp where he lived consisted of a company office, a store, long bunkhouses, and a dining room. The huge bunkhouses at such construction sites were usually made of logs, with floors of rough board. Dozens of men slept in double-deck bunks lining the walls. A large stove occupied the centre of the floor. One man who had inspected these camps wrote that sanitary regulations were often neglected, and that the blankets the men slept in were often dirty. "Vermin are rampant in camps long used," he wrote, "the blankets themselves smell heavy and musty, and, even though one changes underwear every week, it is impossible to keep clean, for lice and nits are in the bedding."[7]

Life in the camp in the off hours was not too bad. The men could read or play cards, and for those who wanted more excitement, there was gambling, opium smoking, arguing, or fighting. Too often there were fights, and Roscoe didn't like the way the other men viewed the activity. "Men are very much like a flock of cockerels and in many cases haven't much more common sense. If there was the slightest affront then a bout of fisticuffs was seen as the only possible settlement and certain conventions were religiously observed. There must be no interference on the part of bystanders even if somebody is killed or in danger of being killed." Roscoe, a small man, found himself the target of a fight one night when a discussion about politics turned into an argument. "I was very badly mauled by a big Liverpool Irishman and there was no interference from anybody. I think they all would have sat idly by and watched him kill me. He was huge and treacherous and it is altogether likely that most if not all were afraid to intervene." Roscoe was badly bruised, but he survived.

After Roscoe had been on the job for more than two months, the company doctor advised him to quit and leave the camp. The doctor feared that either the wet conditions at the construction site or the dynamite fumes in the tunnel might be dangerous to his health in view of the fact that he had been diagnosed as having tuberculosis. So Roscoe and four other workers made plans to head east by hopping on freight trains. After collecting their final pay, they climbed aboard a train that took them a short distance to Laggan, at the summit of the Rockies. They found that there were no freights going out until the next evening so they spent the night in an empty boxcar in the Laggan freight yard, huddled in blankets against the cold. In those days the men who rode the rails were known as "blanket stiffs," because they all carried at least two blankets.

The next morning the men were hungry, but they were afraid to venture out to find food because they knew that the North West Mounted Police were stationed in Laggan. If they were caught on railway property they'd be arrested. During the course of the day one of the men left the boxcar and didn't return. Roscoe and the others, concluding that he had been "pinched," were more cautious than ever. For most of the day the cold men sat in a corner of the freight car, once again huddled together for warmth, waiting for a train heading east. The hunger grew too strong. In the early evening one of them collected some change from the others and, with twenty-six cents in hand, left to buy some food. When he returned with bread and cheese, the group "lit into it like a pack of famished wolves," Roscoe recalled. "If I remember right, we didn't even bother to take the cloth off the outside of the cheese."

Later that evening the men questioned a sympathetic yardman and learned that a freight was to leave soon for Calgary. The yardman even helped them plan their escape. He advised them to wait until the freight train was linked up and had started to move out of the yard. This way they wouldn't be spotted by the brakeman, who walked the length of the train and swung his lantern into each empty car. Roscoe and the other men were to stand between cars on the next track, and when the train began to move they were supposed to swarm into an empty car, staying together for warmth as well as protection against the occasional hostile train man.

Everything seemed to go fine. When the train moved, Roscoe dashed from his hiding place and hurled himself into a car with an open door. Too late he realized he was alone in the boxcar—he had missed the car that the rest had scrambled into. It was cold and dark and he was alone without the protection of the other men.

For a while things didn't seem too bad. It was a glorious, clear moonlit night and Roscoe watched the peaks of the Rockies go by. He thought that anyone who had not seen the Rockies at night in mid-winter had missed one of the great wonders of the world. But gradually he got cold, and finally he had to keep moving around the boxcar in an attempt to warm up. "The jolting of the car gave me a terrific pain in my side and stomach. It was so cold that within a short time I was compelled to remove my lumberman's rubbers and run back and forth to prevent my feet from freezing."

So Roscoe walked and ran the length of the car, always watching the door carefully to make sure it didn't quite close. Grain car doors had a snap lock that sprang into place on the outside, and he had visions of being locked in that car for a week or so until his frozen body would be found when the car was opened to receive grain. "I kept walking, counting the number of steps I had walked." Sometimes in the course of his rambles he went to sleep and would only wake up when he stumbled into the end of the car.

Then the train slowed down, and stopped. Roscoe calculated that he had been travelling for about two hours. He presumed that the train had stopped to take on water. While he was speculating on how much longer he could stand the cold and the jerking motion of the train, he heard footsteps coming along the station platform. He quickly sat down in the corner of the car, pulling his dark grey blankets over himself. The footsteps stopped. "The door was pushed open and a lantern, followed by the face of a 'brakey' thrust in. I could see my finish. But with a laconic 'Hullo,' he retreated, closing the door with a bang."

For a moment Roscoe felt relieved. He wasn't going to be arrested. But then he realized that the car door had closed. The click of the spring bolt had been plainly audible. "I found that I had been fastened into this car and was a prisoner with the mercury away below zero.... The train started. I felt like shrieking and may have done so. I was nearly mad with fright."

When he calmed down, Roscoe decided that the next time the train stopped and the brakeman approached he would make enough noise to call attention to his plight. The important thing was to keep warm. But after walking for an hour or so he got discouraged and lay down in the corner, wrapping his blankets around him. Cold, hungry, and depressed, he dozed off and was lucky he didn't freeze to death on the floor of the boxcar.

Some time later something woke him. When he opened his eyes, he saw a beam of moonlight running across the floor of the boxcar. Startled, he looked up and discovered that the light was coming through an open trap door in the end of the car. He realized he'd be able to get out through the hole. When the train stopped at the next station he crawled out through the small door, opened the side door, and got back in the car before it pulled away again. He was relieved and felt certain that now, even though he was cold and hungry, he would not die. From then on Roscoe remained close enough to the door to make sure it remained unlatched.

After the night-long trip the train pulled into the freight yards in Calgary and a half-frozen Roscoe gathered up his blankets and bag and hustled to the station and into the waiting room. He wasn't even half thawed out when a big policeman came along and asked, "Where you going?" Roscoe said, "Back East." The officer said, "Let's see your ticket." Roscoe couldn't produce one so he was hustled out onto the street.

Instead of trying to sneak back onto the railway lands and catch another train, Roscoe decided to spend a little of his precious money on a hotel room. The cheapest room he could find didn't have a bath. So he crawled immediately into bed, dirty and tired from his near-fatal trip—one of the most frightening experiences of his life.[8]

❧

The next morning Roscoe walked through the streets of Calgary and decided to spend some time there. He found a flop house where for twenty-five cents a night a person could sleep on a narrow cot in a large dormitory. There was also a large lounge where people could sit and read. For $5 he and other unemployed men could buy a restaurant ticket that would provide twenty-one meals. Roscoe got into a routine of sleeping late in the morning, eating a good meal around noontime, and buying a coffee and a couple of doughnuts at supper time. He was able to survive for a time on his railway construction earnings, and he knew that if his situation became hopeless he could call on at least a half-dozen relatives in the East to come to his rescue. But he thought of himself as one of the "down and outs," and once when he recognized a young woman from back home at a meeting he was too embarrassed to speak to her.

Many men who could not get work during the recession that began in 1907 lived on handouts. There were debates about whether it was proper to steal to stay alive if you were starving. Some socialists and radical union leaders strongly defended the right of a man to steal to feed a hungry family if he could get food no other way.

To earn a little money, Roscoe sold copies of *The Eye Opener*, a Calgary weekly published by Bob Edwards. Edwards was a heavy-drinking Scot who used the paper to harass and poke fun at the establishment. Roscoe liked Edwards for his defence of the unemployed. "He had a big heart," Roscoe wrote later, "and Big Business could not force him to condemn hungry and desperate men no matter what they did or said."

Edwards's stinging personal attacks on politicians and businessmen occasionally ended in lawsuits against *The Eye Opener*. In 1908 the paper published a long letter in which a woman accused Conservative leader Robert L. Borden of being the father of her daughter's illegitimate child. When the paper was sold in Borden's home constituency in Halifax just before election time, Borden sued the paper's distributor for libel. No charge was laid against *The Eye Opener*. The paper was also famous for its scurrilous "Society Notes":

> Society Note: It is unlikely that T.B. Mulligan will run for the council, as announced in the press. Mr. Mulligan does not get out of jail until December 20, too late to file his nomination papers.[1]

Edwards's sharp barbs and off-colour stories meant that insults were occasionally thrown at Roscoe and other newspaper hawkers as they stood on street corners selling *The Eye Opener*.

It seemed to Roscoe as though the streets of Calgary in December 1907 were full of angry unemployed men, socialists, and radical unionists. As he had done when he lived in Maine, Roscoe quickly became involved in radical activities. He began to promote socialist literature published by the Charles K. Kerr Co., a co-operative publishing venture with headquarters in Chicago, which included the U.S. radical labour-leader William "Big Bill" Haywood among its editors. Roscoe was one of hundreds of people throughout North America to buy what were called "shares" in the company. The so-called "shares" didn't include voting rights in the co-operative, but they did give supporters the privilege of buying the company's books at cost. The company sold the shares to get capital to publish more books.[10] Roscoe took a special interest in the company and distributed free pamphlets describing its literature. The company also published *The International Socialist Review*, a monthly magazine that would later become one of the largest circulation magazines of the left in North America.

One day when Roscoe was passing out socialist pamphlets, a man came up and introduced himself as Jack Leheney, an Alberta union activist. Roscoe and Leheney became friends and Leheney arranged for Roscoe to attend the first annual convention of the Alberta Trades and Labour Congress. The convention was Roscoe's first contact with organized labour. The issue of the day was labour's involvement in politics—whether a new labour party should be organized or whether workers would support an already-existing socialist party.

Leheney proposed that Alberta labour endorse the Socialist Party of Canada, which advocated replacing capitalism with a co-operative society. The Socialist Party had been formed in 1904 from an already existing socialist party in British Columbia and a socialist group in Ontario, and it was moving into other provinces and trying to build support among existing labour organizations. Roscoe looked on with excitement as Bill Davidson, a former B.C. Socialist Party MPP and an officer in the Western Federation of Miners, told the Trades and Labour Congress that the working class must question the right of capitalists to own the means of production.[11] The convention voted thirty-seven to nine to adopt the policies of the Socialist Party and pledged to raise funds that would be used to promote socialism.

For Roscoe, the convention was a great morale booster. "I was full of beautiful dreams of the future society," he wrote later. Back in the Maritimes hardly anyone knew much about socialism, but in the West an exciting new political movement was taking shape. Many Western workers, especially B.C. miners, had become radicalized by the arrival of full-fledged corporate capitalism in the hard-rock mining industry that stretched from the Mexican border northward into the Canadian Rockies.

In the United States in the 1890s, large corporations solicited the support of federal authorities to crush revolts by miners who were seeking better pay and better working conditions.[12] Working conditions in the mines were particularly oppressive. Fearful that safety regulations and labour standards would discourage investment, governments were reluctant to introduce protective legislation. Although workers lived in a brutally exploitative environment, they were also aware of the conspicuous consumption of the Edwardian era—galas, automobiles, stately houses—and these conditions were conducive to the growth of radicalism.[13]

In places such as the Coeur d'Alenes, Leadville, and Cripple Creek, federal troops often engaged in open battle with armed miners, routing the workers and throwing their leaders in jail. To combat the combined forces of the corporations and government, the miners had formed the Western Federation of Miners (WFM), which soon embraced Marxism and became a home for radical socialists. The WFM had organized miners in British Columbia and Alberta by the turn of the century but, facing opposition from the mainstream American Federation of Labor, was in decline by the time of the Calgary convention. Roscoe met men at the convention who belonged to the WFM, and his thinking was greatly influenced by their stories of how corporations and politicians were out to destroy the radical labour movement.

Having lost several struggles, the Western Federation of Miners soon realized that it could not compete with the combined powers of corporations and government through the traditional labour-management bargaining relationship. It believed in the necessity of making a fundamental change in the basic structure of society. In 1905 WFM leaders such as Big Bill Haywood formed one big union—the Industrial Workers of the World (IWW)—which would organize all workers into industrial groups. Haywood envisioned the IWW growing so large that it would eventually topple and replace capitalism with a co-operative society.

Roscoe liked the IWW's philosophy—he too was becoming convinced of the need for a new social order to replace the old. He started to read the Socialist Party newspaper, *The Western Clarion*, and was soon familiar with the party's Marxist doctrine. Like the IWW, the Socialist Party believed that capitalism could not be reformed and that attempts to co-operate with it had no place in the class struggle. The party also opposed the existence of a trade union movement because it believed that trade unions might benefit some workers in the short term but would fail all workers in the long run. According to the party, political action based along class lines was the only means by which working people could destroy the wage system that oppressed them.[14]

For Roscoe, the need for change could be seen on the faces of the hundreds of unemployed men flocking into Calgary. Many of them were lost souls, some from as far away as Britain. They had no work and they were hungry. These unemployed, who may have numbered four thousand, were angry that neither business nor the city would do anything to help them. While a lot of money was going into the construction of new exhibition buildings for the city, no jobs were being given to single, unemployed men who were not from the Calgary area.

Roscoe's exposure to socialism during these few weeks in Calgary caused him to change his attitude toward his own poverty and unemployment. He was aware of this change on Christmas day, when he was starting to run out of money and had no place to go for dinner. As a last resort Roscoe and a few other unemployed men decided to go to the Salvation Army, where they were able to sit down to a hearty dinner. Roscoe later recalled how most of the men at the dinner were ashamed of their poverty. "With scarcely a single exception they sat with their heads down, trying to hide their faces in their plates." Preparing the meal were several prominent Calgary women dressed in silks and broadcloth. Roscoe believed that the women saw the unemployed men as a novelty, and that they "crowded round to see the human animals eat at the Salvation Army dinner."

While most of the unemployed men were embarrassed about their poverty, Roscoe and the few socialists at the dinner had a different attitude. "We held our heads as high as though we were dining at the Waldorf Astoria or St. Regis.... We knew that we were eating food which had been stolen from us. We were just getting what we had produced. And as we ate and enjoyed the 'charity' of the masters, we dreamt of the day when we will take the earth and enjoy all the good things as our right."[15]

Before long Roscoe and Fred Hyatt, a young, aggressive British immigrant, were organizing the unemployed in Calgary. Hyatt was one of hundreds of unemployed men from England in the West who were angry because they felt they had been misled by Canadian immigration authorities about the opportunities available to them in Canada. The initial meeting of the unemployed, held in early January 1908, drew so many people that the streets were blocked by a traffic jam. Many protesters were jostled by police but there were no arrests. The unemployed movement was not well received by either city officials or the press.

"The Calgary newspapers were very sarcastic over the blundering speeches many of us made and hinted at dire things in the past record of those active in the movement to get jobs," Roscoe wrote later, referring to a claim by *The Daily Herald* that Hyatt had left a job that paid $2.75 a day. "The newspapers in general in those days took the position that the unemployed men were perfectly, naturally and inevitably expendable and that 'progress' must go on no matter what it did to the individual. Or, alternatively, the unemployed were a bunch of shiftless goons who wouldn't take a job if it was offered them and were demanding that society support them in idleness and affluence."

Hyatt continued to work with the unemployed, staging protests and leading delegations to various government offices to demand better living conditions for the unemployed. He soon became the Socialist Party's organizer in Calgary.

By late January 1908 Roscoe had little money left, and with no chance of finding work in Calgary he decided to go home. His brief stay in Calgary had been worth it: He had made contact with active radical socialists and learned how they went about their work of organizing opposition to capitalism. He had become emotionally wrapped up in a political movement that demanded justice for all workers. He dreamed of the day when a strong brand of socialism would spread to the poor people of New Brunswick, who he felt so badly needed to stand up for their rights.

FOUR

Socialism for the Maritimes

In many ways the depression conditions of Calgary in the winter of 1907-08 reminded Roscoe Fillmore of the hardships that had been a norm in New Brunswick for as long as he could remember. Men weren't clamouring in the streets of Albert for industrial jobs like they were in Calgary, but the poverty that Roscoe saw among the poor farming families when he returned home had been there for generations.

In Albert Roscoe looked up his old friends and began talking about socialism to any of them who would listen. He thought it was senseless that people not just in New Brunswick but across North America had gone hungry during the winter. In an article published a year or so later in the Canadian socialist newspaper, *Cotton's Weekly*, Roscoe wrote: "The depression has shown the workers as never before the absurdity and cruelty of a society in which millions are starved because too much food and clothing have been produced. Warehouses filled to overflowing and the producers starving are spectacles which should cause every intelligent man to think."[1]

For Roscoe the solution was not more productivity, which was what some business people and politicians were advocating. Instead the solution was socialist political action. Socialism seemed to be on the upswing: In recent elections around the world, nearly eight million votes had been cast for socialist candidates, with 489 people elected in nineteen countries.[2] It seemed it wouldn't be long before socialism would appeal to people everywhere.

Roscoe began to order as many socialist newspapers, magazines, and books as he could afford, and the small bedroom he shared with his brother

Frank was soon full of political material.[3] Night after night until well past midnight, Roscoe sat at the kitchen table reading the classics of socialism, including Marx and Engels. He wrote to the Vancouver headquarters of the Socialist Party of Canada and asked them to send him information about how to form a chapter in Albert.

Roscoe's return to Albert meant that his stepmother Selina had a family of nine to feed. Roscoe's sister Ellida was nineteen and well on her way toward a teaching degree. Frank was fifteen and doing well in school. Roscoe's half-sisters Clara and Mabel were eight and five years old. Also at home were two of Selina's children from her previous marriage.

According to Clara, Roscoe didn't attempt to impose his political beliefs on his family. Much later she recalled him as a "very easy going, soft-spoken man." She said, "He was outgoing, and would sit and talk for hours about all kinds of things, but he wouldn't try to preach to you about his socialism." The family noticed how serious, almost solemn, he had become. Clara thought he had always seemed old for his years: "He was born old."[4]

Roscoe's worst characteristic was a bad temper, inherited from both his grandmother and his father. If a machine at the nursery wouldn't work, he'd launch into an abusive verbal attack. "He'd get awful mad, but we'd start laughing at him and the first thing you'd know he'd be laughing along with us," Clara said. With the exception of his hot temper, Roscoe seemed to try to keep his emotions in check—a trait he may have also inherited from Elizabeth.

Roscoe's personal time was often spent in solitary activities that gave him time to think. He enjoyed reading, fishing, and walking in the woods. He had taken up smoking. Roscoe had learned to drink when he lived in Portland, but no liquor was tolerated in his father's house. His stepmother had an especially strong dislike of liquor: Later, when she lived in Detroit with her daughter, Selina always did the dishes after dinner but refused to wash any glasses in which liquor had been served.

At the nursery, Willard and his brother Arthur were extremely busy growing and selling fruit trees throughout the region. Roscoe, unable to find other work, joined them. He liked the work and was interested in the flowers, shrubs, and trees native to the New Brunswick countryside and how they compared to those he had seen in the nursery in upstate New York.

After Roscoe received the information he wanted from the Socialist Party, his cousin Clarence Hoar, who worked at the local branch of the Bank of New Brunswick, helped him with the groundwork needed to set up a new chapter. They showed all their friends copies of *The Western Clarion*, the Socialist Party's newspaper, and the other socialist literature. By March 1908 they had the minimal five names required for a chapter. Two other members who helped with the work were farmer Claude Davidson and Robert Tingley, a former schoolmate of Roscoe's. The charter cost five dollars, and individual members were required to pay dues of about twenty-five cents a month. Roscoe sent the application off to Socialist Party headquarters in Vancouver, saying that he hoped to have an additional five members very soon.

Roscoe was attracted to the Socialist Party partly because most of its writers and lecturers were workers with little formal education, like himself. The leaders of the movement were self-instructed in Marxism and related topics. They worked at their jobs full-time and taught, studied and organized in their spare time. The party's strongest base in the country was in British Columbia, where it had elected members to the legislature, and Roscoe was encouraged by the progress the party was making as it moved eastward into Manitoba and Ontario.

But in the Maritimes in 1908 only a few people knew anything about socialism. The first socialist club in the region had been established ten years earlier in Halifax, eventually becoming a section of the Socialist Labor Party, which had its headquarters in New York. The group published its own monthly newspaper, *The Cause of Labor*, and held open-air public meetings in Halifax parks and on the street. But the Halifax group ceased to operate after two of its key members moved to Cape Breton, where they set up socialist clubs in Glace Bay (1904) and Sydney Mines (1905) among the radical immigrant coal miners.

The earliest prominent socialists in New Brunswick were Martin Butler and H.H. Stuart. Butler, who had lost an arm in an industrial accident and earned his living as a vagabond, peddler, and sometimes-journalist, was thirty-two years old in 1890 when he launched *Butler's Journal*, a small, opinionated monthly published in Fredericton. Butler frequently wrote at length about democracy and Canadian independence and gradually accepted the ideas of socialism after meeting H.H. Stuart. Stuart was fifteen when he took up a career as a printer in Fredericton in 1888, later became a school teacher, and wrote about poverty and social justice for *Butler's Journal*. He appears to have adopted a socialist philosophy by 1897, when he wrote an article on socialism in the U.S. newspaper, *Appeal To Reason*. Stuart endorsed the Socialist Labor Party.[5]

In 1902 Stuart and Butler established the province's first socialist group in Fredericton—branch No. 67 of the Canadian Socialist League—with a goal of running candidates for political office. This group took a stronger position than any before it, urging workers to act as a class in their struggle against capitalism.[6] In 1905, after the formation of the Socialist Party of Canada, Stuart and Butler set up a party chapter in Fredericton. Two years later Stuart went to Cape Breton to help establish a local of thirty-seven members. At about the time Roscoe arrived home from Calgary, a local of forty members was set up at McAdam Junction, New Brunswick, which was the centre of the CPR system east of Montreal. Socialist Robert Scott was active in Halifax, distributing a large volume of socialist literature, but a Socialist Party local had not been formed there.

❧

Roscoe and Clarence Hoar faced even more resistance in getting people in Albert interested in socialism than they had expected. In fact, it would have been hard to find an area anywhere in Canada more resistant to radical political ideas than Albert County. People were individualistic and proud, and they thought that the old-fashioned, conservative ways were best. The villages of Albert and Riverside voted overwhelmingly for the Conservative Party. In six provincial elections held between 1900 and 1925, the Tories carried the county five times, with an absolute majority in each case.[7] More than 70 per cent of the people in the county were fundamentalist Baptist, and in such conservative surroundings the socialists had little success. Roscoe reported to Vancouver: "We are having a hard time to keep things running just at present. Many of the people would starve us out if it lay within their power."[8] Some people became openly hostile and would cross the street in the village just to avoid speaking to Roscoe.

The socialists rented a building on Main Street in Albert, and Roscoe was shocked by the response to their new headquarters. One evening in June they were looking over the building and measuring the windows for curtains when they heard a crash against the door. A group of young men and a woman began throwing stones at the building. A window was broken, but no one was injured. Roscoe later found out that the stones had been thrown by the sons of "eminently respectable citizens" of the village. Despite the incident, the new office was soon open. They began to hold regular meetings and established a small library where anyone could come in and read the latest in socialist literature.

When Roscoe gave his first public speech at the new socialist offices he didn't present a very imposing figure. He was thin and youthful—about five-foot-six and 130 pounds—and usually dressed in work clothes. He had a gaunt, handsome face, with wide-set, penetrating blue-grey eyes, a broad nose, strong jaw, and a stern-looking mouth like his Grandmother Elizabeth's. His short, light-brown hair was parted on the side and often carelessly tumbled onto his forehead. He was a bit awkward and shy in public and spoke in an earnest, halting manner.

Perhaps the most influential person opposed to socialism in Albert was

Rev. W.A. Snelling, minister of the Baptist Church. Rev. Snelling was president of the local Purity League, an organization that, among its various activities, attempted to prevent corrupt election practices. Roscoe, who believed the organization's real purpose was to try to strengthen the position of the Conservative candidate, attended one of its meetings and bluntly gave those in attendance his opinions on the causes of political corruption. After that meeting Rev. Snelling never missed an opportunity to criticize socialism. Once when Rev. Snelling associated socialism with anarchy, Roscoe challenged the minister to a public debate. The minister declined the challenge.

Roscoe's conflict with Rev. Snelling was an embarrassment for Willard Fillmore, who was one of Albert's most active Baptists and a Deacon of the church. Willard was often responsible for collecting money from parishioners to pay church bills and occasionally represented the church at district meetings throughout the province. Much to Roscoe's irritation, Rev. Snelling and other Baptist ministers were frequent dinner guests at the Fillmore home. Willard used to say that the Fillmores would have been rich if they hadn't had so many Baptist ministers to dinner. Roscoe's half-sisters were also active in the church, so they were less than pleased when Roscoe's socialism became a topic of discussion in the village. Mabel said that Roscoe's activities "hurt us, because everyone in the neighbourhood was either Catholic, or Baptist, or Methodist or Anglican, and they were great church people and they looked down on Ross."[9]

Willard soon began to dislike Roscoe's politics, and the two of them would engage in heated arguments. Selina also didn't approve of Roscoe's new ideas but kept her opinions to herself, frequently criticizing Willard for needling Roscoe. But most of the time Roscoe had a good relationship with his father. If Roscoe ever needed money, or if he wanted advice, he could always go to Willard. At times Willard also showed a certain curiosity and openness about socialism—once he said he'd like to go with Roscoe to Moncton to hear a prominent Canadian socialist speak—and Willard and Selina willingly opened their home to any of Roscoe's socialist friends who needed a place to stay.

Local business owners opposed the socialists. Roscoe felt that many people interested in socialism dared not join the party because they feared they might not get employment in the village. Roscoe had been refused a job by one of the local employers, and one of the charter members had quit because his affiliation with the socialists had hurt his business. Roscoe wrote to H.H. Stuart, now living in Newcastle: "Isn't it hell that a man should be compelled to be a hypocrite in order to obtain his bread and butter."[10]

Roscoe found that villagers were changing their attitude toward him. "Our family had a good reputation as honest, hard working people," he wrote later. "I personally had a good name as trustworthy, but I noticed that the storekeepers now began to watch when I came into the stores. I believed in an unpopular doctrine—presto, I had become a scoundrel and a disgrace to my deserving parents! Poor old Albert needed new ideas as much or more than most places, but it was constitutionally unable to accept them."

Roscoe did what he could to try to reverse public opinion. With hundreds of thousands of people around the world joining the socialist movement, he felt it would be just a matter of time before working-class people in New Brunswick would become interested. One step Roscoe took was to write a thousand-word essay on socialism for *The Albert Journal*, the county's only weekly newspaper. The essay was published as the main feature on page one under the headline: "A SOCIALIST SPEAKS. Roscoe A. Fillmore, of Albert, Writes Entertainingly of Socialism." Trying to correct what he saw as popular misconceptions, Roscoe argued that a socialist government would own property in the same way that present governments owned the Intercolonial Railway, the post office or, as in some provinces, the telephone system. He said that socialism should not interfere with a person's religious beliefs and pointed out that "some of the brightest minds in the pulpit" in Canada and Europe were socialists.[11] The article did little to change people's minds about socialism.

Only a few months after its founding, Roscoe's little socialist band faced a serious setback. Clarence Hoar, who had joined Roscoe in writing for the socialist press and who worked hard at spreading socialism, was told by the Bank of New Brunswick that he must either resign from the Socialist Party or leave his job at the bank. Even though Clarence had no prospect of getting another job, and despite his fear of being blacklisted, he resigned from his job. He couldn't find another job in Albert County so he moved with his family to Portland, where he continued to be a strong socialist supporter, joining the Socialist Party of Maine and working on the socialist paper, *The Issue*. He maintained his membership in the Albert local.

There were other problems in Albert. Another party member quit the group after an ultimatum from his employer, and two members were expelled from the local for voting for mainstream party candidates in the 1908 federal election.

❧

Despite the difficulties in Albert, Roscoe was eager to try to build socialism in other parts of the Maritimes. He placed an ad in *The Saint John Daily Sun* asking socialists in the Maritimes to get in touch with him and was surprised when several people responded. One of the most promising replies was from the family of William Mushkat, a Moncton dry goods merchant. Roscoe went to meet the Mushkats and learned that several members of the family were committed socialists who had experienced political oppression in their Polish homeland. He went to visit them often after that and was particularly impressed by William's daughter Sophie and her friend Fanny Levy. "The girls are thoroughly well read," Roscoe wrote to Stuart. "They beat any women I have ever met in the movement. No silly talk about hats, ice cream, etc. They are the genuine article and no mistake."[12]

In 1909 Roscoe threw himself into the unpaid position of Maritime organizer for the Socialist Party. He continued to work at his father's nursery but spent every other available minute working for the party, especially writing letters to prospective Socialist Party members in small towns throughout the Maritimes. He was supported in this work by H.H. Stuart, who at thirty-five years old was more experienced in politics than Roscoe, who was twenty-one. Stuart had played a prominent role in the formation of the teachers' association and was active in the temperance movement, but he had already paid a price for his interest in radical politics: He had been dismissed from a teaching position at Fredericton Junction for circulating socialist materials and for talking about socialism in public places.

Their work was also helped by a few active publications. Until 1909 the Socialist Party's main outlet for information was *The Western Clarion*, published in Vancouver. The appearance of a second socialist newspaper that year was a boost for the fledgling party, especially in the Maritimes. *Cotton's Weekly* was founded and edited by W.U. Cotton out of Cowansville, Quebec, about forty kilometres east of Montreal. Modelled on the successful U.S. paper *Appeal to Reason*, it soon had more than a thousand subscribers in the Maritimes.

Four small Maritime publications also showed an interest in socialism. *Butler's Journal* supported the cause, occasionally featuring articles by H.H. Stuart. From 1907 until 1909, Stuart himself was editor of *The Union Advocate*, a small weekly in Newcastle, New Brunswick. Once Stuart left the paper it became hostile to his ideas. In Moncton, Bruce MacDougall published an aggressive weekly called *Free Speech*, but the paper was suspended in 1909 after MacDougall published an attack on several prominent Saint John citizens and was convicted of libel. *The Eastern Labor News*, published in Moncton and endorsed by the Moncton Trades and Labor Council, was closer in its politics to the traditional labour movement but sometimes covered the activities of the socialists. Mainstream newspapers such as *The Halifax Herald* and *The Moncton Times* occasionally covered the activities of the socialist movement but usually put a negative face on socialism.

Roscoe and the other radicals wrote dozens of articles in the socialist press to try to attract members to their movement. "My reader, you cannot afford to neglect the study of socialism," Roscoe wrote in *Cotton's Weekly* in March 1909. "Drop the capitalist paper if you haven't time to read both and read the paper which is edited and supported by your class."[13]

❧

Roscoe and H.H. Stuart wanted to attract new members to the party and establish new locals throughout the Maritimes. One way of doing this, they thought, was to get a prominent socialist to tour the region. British Labour MP Kier Hardie had been in the Maritimes during a Canadian tour in 1907 and had generated a lot of public interest. The Maritimers were offended that E.T. Kingsley, the intellectual leader of the Socialist Party, had toured from Vancouver to Montreal in 1908 but had not come to the Maritimes as expected.

The Maritimers put pressure on the Dominion Executive to send someone to tour the region. In April 1909 the party announced that Wilfred Gribble of Toronto, who had just been named the party's national organizer, would travel through the Maritimes for about two months. Gribble, a carpenter by trade, was known for his strong speaking abilities.

Roscoe took on the task of organizing Gribble's tour. He drew up the itinerary and made sure that socialists in the various towns arranged publicity and places to speak. In most towns there were no active socialists, so Roscoe got a list of the *Cotton's Weekly* subscribers and wrote many of them asking for their help. While the national executive paid part of the cost of Gribble's tour, Roscoe established a special fund to raise the rest of the money. He appealed for help in *The Western Clarion:* "New Brunswick and Nova Scotia are ripe for socialism. There is discontent everywhere and we must lead that discontent into intelligent channels. In order to do this we must have money, money, money!"[14] Socialists from across the country donated more than $115 toward the cost of the tour.

Gribble arrived in the Maritimes at the beginning of May. He went to Newcastle, where Stuart had recently organized a party local, and then moved on to Albert. Gribble was pleased to find a socialist chapter in such a small village as Albert. He reported in *Cotton's Weekly*: "They have a large sign up, 'Socialist Headquarters' in letters that can be read a quarter of a mile away and this in a tiny village. Think of this, some of you comrades in places where there are large locals.... Think of socialist headquarters in a village where one of the 'leading' citizens talks of driving out the socialists with shotguns, where a 'Christian' lady tears down socialist bills, and a village belle breaks the windows of the socialist headquarters."[15]

Roscoe thought Gribble was just what the Maritime socialists needed. Gribble, tall and imposing with a booming voice, turned out to be an excellent speaker. Sixty people came out to hear his first public speech in Albert and forty showed up at Socialist Headquarters to talk with him. It seemed that no one would be able to avoid thinking about the need for socialism after listening to the party organizer. Gribble explained that labour was a commodity bought and sold in the market just the same as any other commodity. Industry was interested in maintaining a surplus of labour so it would be cheap to buy. Improvements in machinery and the growth of powerful trusts were putting more and more workers out of their jobs, increasing unemployment. The only alternative workers had was to use their political power to seize the means of production and produce everything for their own use instead of for the profits of a master class. For weeks afterwards the Albert socialists benefited from the visit. "I have had good meetings in our hall every night this week, four in all," Roscoe wrote to Stuart in June. "The fellows come in and stay all of the evening and listen, talk and discuss the question. I never felt so happy in all my life."[16]

Gribble's travels through the Maritimes included stops in all of the large centres as well as many small towns and villages such as McAdam Junction in New Brunswick and Weymouth, North Range, and Caledonia in Nova Scotia. Roscoe tried unsuccessfully to arrange a speaking engagement for Gribble in Prince Edward Island.

Although Roscoe and Stuart were pleased by Gribble's reception, there were still problems. The worst was a lack of money. The dues the Socialist Party collected from its few members barely amounted to what was needed to pay for a little travel and some literature. Roscoe and Stuart kept in touch with their locals by letter, not the most effective method of rallying people to the socialist call. Several times Roscoe and Stuart dipped into their own pockets to pay travelling expenses or to hire a hall. On occasion Roscoe received expense money from the party.

In Cape Breton, Gribble was surprised to discover a group of committed socialists who would form the basis of a radical movement that would function for years after. Most of the men were from Scotland or other parts of Europe and had a solid understanding of socialism: people such as J.B. McLachlan, the radical mineworkers leader who had learned his politics from Scottish socialist Kier Hardie; A.A. McLeod, later a member of the Ontario legislature; D.N. Brodie, a local printer who would become chairman of the Maritime executive of the Socialist Party; and Alex McKinnon, who Gribble said had a knack of simplifying complicated subjects.

Gribble arrived in Cape Breton just before the 1909 strike by the miners, who were fighting to have the United Mine Workers recognized as their union. Gribble, who believed that unions were incapable of defending workers against capitalism, made himself unpopular with some of the miners by arguing that their strike would fail. Still, he found enthusiastic audiences. At Sydney Mines the audience prevailed upon him to speak until his voice squeaked. When he couldn't go on, they started to sing *The Red Flag*. That evening Gribble collected twenty-five membership applications. After the meeting Gribble and a few other socialists went to McLachlan's home, where they talked economics and sang until midnight. On Gribble's recommendation, the executive of the Maritime party was set up in Cape Breton.

In Glace Bay, John L. Lewis, the much tamer leader of the United Mine Workers of America, was also in town. Glace Bay was a stronghold of the Catholic Church, and Gribble sat in church as a priest criticized the socialist movement and proclaimed that Lewis was a better friend to the working man. After the sermon, an incensed Gribble took over a vacant lot across from the church and started delivering a speech that drew so much attention it caused a traffic tie-up. Police were called, and when Gribble refused to stop speaking he and a handful of his socialist supporters were taken to the police station. Gribble finally agreed not to speak any more that evening.

Gribble's tour was scheduled to end in early July, but after Roscoe made an appeal in *The Western Clarion* for socialists to donate more money, the tour was extended. On this part of the tour Gribble ran into some resistance in Stellarton, Nova Scotia. He delivered speeches two nights in a row to well-attended outdoor meetings, but by the third night some of the businessmen of the community had asked their workers not to attend. The attendance that night was smaller. To make matters even more difficult, members of the Salvation Army, which was strongly opposed to socialism, positioned themselves nearby. Gribble wrote that the Army showed its Christian spirit by "thumping the drum and howling something like 'Oh, you must wear a collar and a tie, Or you won't go to heaven when you die.'"[17]

While Gribble was touring Cape Breton, Roscoe did some campaigning of his own in Moncton, where he had his first brush with the law. Roscoe was holding forth on a vacant lot when a policeman told him that the chief of police wanted to see him at city hall. After a bit of a scuffle, Roscoe went along and met the police chief, who told him he was obstructing traffic on Main Street and would have to stop speaking. Roscoe said he wouldn't speak again that night, adding that he intended to make another speech the following night. The police chief said this would be possible only if the socialists could get permission from the lot owner to use the property.

When Roscoe woke up the next morning he was surprised to read in the

Moncton paper that the city would have no more socialist street meetings. More determined than ever, Roscoe tried to get permission to use the vacant lot but found that the owner lived in Saint John and couldn't be reached. That evening Roscoe, believing strongly that anyone should be allowed to address the public in a public place, began speaking once more at the vacant lot. This time Clarence Hoar, Sophie Mushkat, and other socialists were on hand to encourage people to move along so the street wouldn't be blocked. The police didn't interfere, although the chief again asked Roscoe to come and see him, at which point the police chief told Roscoe not to speak on the street again or he would be arrested and charged. Roscoe told the chief, "That is your duty, mine is to speak." Somewhat flabbergasted, the chief looked at Roscoe and said, "I think we can find a place for you to speak." He came up with a vacant lot and there was no further trouble.[18]

When Gribble came to Moncton, labour leaders denied him permission to speak at a huge Labour Day rally. At the close of the rally, Gribble and Roscoe got up in the audience, spoke for a half-hour, and announced that a street meeting would be held that evening. That night Clarence Hoar, visiting from Portland, delivered his first public speech on socialism, to the delight of Roscoe and Gribble. The next day's news was not so good: *The Moncton Transcript* said Gribble had advocated the use of violence in Canada to spread socialism. An angry Gribble wrote a letter to the editor stating that what he had said was that he supported the violent actions of the revolutionaries in Russia who were defending themselves against the persecution of a brutal regime. In Canada, he said, the party advocated the use of the ballot to overturn capitalism.

Gribble's long tour of the Maritimes ended in late October with a two-week stay at the Fillmore home in Albert. Gribble went fishing and hunted bear. The Albert chapter, which had already experienced a roller-coaster existence, was again facing serious opposition. Rev. Snelling was leading a group of people in pressuring workers to have nothing to do with the socialists. "Albert is the cussedest little burg you ever heard of," Roscoe wrote to D.G. McKenzie at the Socialist Party headquarters in Vancouver. "As fast as we get a fellow roped in, he finds the place so unhealthy that he has to move out. Two more comrades expect to move next month and I'll be alone again. It makes me mad. Talk about bombs! I'd like to throw a whole gross of them into this place. Prayer meetings and revivals are the whole rage the year round. The only interesting thing appears to be getting converted, lapsing into sin and then re-conversion again about twice a year."[19] Roscoe soon gave up on his goal of keeping a socialist local operating in Albert.

Despite the setback in Albert, Gribble's tour was a marked success for Roscoe and the Socialist Party. The trip had been planned to last for about six weeks, but the party's national organizer had spent five and a half months in the Maritimes and provided the impetus for new locals in Springhill, New Glasgow, Sydney, Sydney Mines, Dominion No. 6, Westville, Moncton, Saint John, and Halifax. There were now fifteen locals in the region. In the towns and villages of the Maritimes their work had also identified committed socialists who would continue to work to build up the party.

Party membership had reached about three hundred in the region. As historians David Frank and Nolan Reilly have pointed out, at a time when the region included about 13 per cent of the Canadian population, Maritimers accounted for a respectable 10 per cent of the membership of the Socialist Party of Canada.[20] A fund was started to hire a full-time party organizer for the Maritimes, and H.H. Stuart promoted the idea of holding an annual convention in the region.

The socialist press was making its presence felt in the Maritimes. *Cotton's Weekly*, the most popular socialist paper in the country, had more than 1,300 subscribers in the region by December 1910. Of these, more than 1,100 were in Nova Scotia, principally in Cumberland County (553), Cape Breton (334), and Halifax (99).[21]

It seemed to the socialists that they were finally getting a footing in the region. H.H. Stuart, the pioneer of the Maritime movement, was pleased that after more than ten years of educational work, carried on by a few scattered comrades at great inconvenience and, in several cases, personal loss, socialism was now firmly implanted in New Brunswick and Nova Scotia. Roscoe felt sure that socialism would win. It was just a matter of time.

FIVE

The Jail Birds of Liberty

In autumn 1909, when he was still working on the final stage of Wilfred Gribble's tour, Roscoe Fillmore found himself with yet another visitor on his hands. Big Bill Haywood, the man who had helped found the Industrial Workers of the World (IWW), was coming to Canada. Haywood was one of the men Roscoe looked to as a leader of the socialist revolution.

Haywood was getting a lot of press coverage in North America in 1909, thanks largely to a highly publicized trial. Haywood and two other officials of the Western Federation of Miners had been tried for allegedly plotting the assassination of the former governor of Idaho, Frank Steunenberg, who had opposed the union's organizing efforts. Steunenberg had died within seconds after being hit by a bomb triggered just as he opened the front gate of his house. A drifter questioned by police had said that Haywood and the others had plotted a revenge murder.

Defended in the case by the legendary Clarence Darrow, Haywood was acquitted—but not before spending more than a year in jail waiting for his trial. When he was finally released Charles Kerr and Co. decided to take advantage of his notoriety by commissioning him to tour on behalf of *The International Socialist Review*. Haywood would give speeches and promote socialism and the price of admission would include a subscription to the magazine.

In the late summer of 1909, Haywood agreed to include a tour across Canada as part of his trip. The Socialist Party of Canada was helping to co-ordinate Haywood's visit and asked Roscoe to organize the Maritime leg of the tour.

When Haywood arrived in Saint John in early November, the IWW was in the middle of an important fight for free speech. One of the main battlegrounds was Spokane, Washington, where city officials were trying to bar its organizers from speaking on the street, mainly because they castigated religion and effectively aroused the unemployed and poorly paid workers. At the time the union wasn't permitted to speak to workers at their place of employment, so one of the few ways it had to reach migrant timber workers and harvesters was to have members stand on street corners in frontier towns and give speeches about the benefits of joining the organization. While Spokane officials were responding to this strategy by busily arresting IWW members—on one occasion there were so many IWW people in jail in Spokane that there was literally no place to sit down—they were not interfering at all with the public speaking rights of the Salvation Army, the IWW's main rival.[1]

Roscoe met Haywood at the Saint John train station, but owing to some mix-up it turned out that no meeting had been scheduled for that evening. Embarrassed, Roscoe and the other local socialists—there were seventeen active members—put Haywood up in a hotel. They'd lost the opportunity to place Haywood in front of a large audience in a hall, but the small group quickly decided they could still hold a street meeting that same night, right after a Salvation Army religious meeting. After the Salvation Army had finished, with a crowd already milling about near the centre of the city at King Square, Roscoe stepped out and shouted "Fellow workers!" The meeting was on. Roscoe announced that Haywood would soon be speaking. But when Roscoe launched into a speech about politics and the economy, Haywood wandered away with one of the local members of the Socialist Party.

The next day's *Daily Telegraph* described what followed: "Fillmore stepped to the curb and commenced to harangue a rapidly growing crowd on the tenets of his creed. Police Officer Henery loomed up in the offing but did not at first interfere with the orator. 'I thought he was advertising some meeting to be held by the Salvation Army.' After a minute, however, when Fillmore launched an impassioned appeal in favour of the equal division of property and other socialistic statements, the policeman thought it was time to take notice."[2] By that time there was a major traffic jam all around King Square. According to the newspaper, Officer Henery told Roscoe, "You must move on, you can't address a meeting here."

"I don't know about that," Roscoe replied. "I have the right to speak."

"Well, if you don't move now when you're told I'll arrest you," the policeman warned.

"I am going to stay here," Roscoe said. The reporter for *The Daily Telegraph*, surprised by Roscoe's behaviour, speculated that Roscoe might have been particularly zealous because of the presence in town of Haywood, a champion of the U.S. free speech movement. But Roscoe's stand was no different than the one he had taken in Moncton a month or so earlier when the police chief had tried to stop him from speaking in public.

Haywood still hadn't arrived back at the square. Policeman Henery and a detective arrested Roscoe and took him to the police station. The newspapers said a search revealed that Roscoe had several socialist publications in his possession. Roscoe was charged under the Dominion Vagrancy Act with obstructing a public walkway and refusing to leave the area when ordered to do so by police.

When Haywood appeared and was told what had happened, he telephoned the police station and asked whether Roscoe wanted bail. Roscoe refused the offer and spent the night in jail, which turned out to be an unpleasant experience. He was locked in a six-by-eight-foot cell that contained a narrow plank for a bed, and he got little sleep because of noise caused by some neighbouring drunks. He also experienced claustrophobia for the first time in his life, and this scared him. He found that many of the men in jail were being forced into work as part of a chain gang. "A heavy ball and chain were attached to one ankle to reduce the chances for escape," Roscoe wrote later. "I didn't fancy this experience but bolstered up my courage before morning to face even this."

The next morning Roscoe's arrest was front-page news in Maritime newspapers. *Cotton's Weekly* said in an editorial that his arrest showed the class character of the police system. "Speech is free on the streets so long as the sentiments expressed are satisfactory to the present system of government."[3] Haywood was scheduled to leave Saint John early in the morning for Springhill, where he was to speak that night. Before he left, Haywood told *The Daily Telegraph* that it was not unusual for a socialist to be arrested for obstructing the streets and refusing to move on. He said that the Saint John local of the Socialist Party would make sure that Roscoe got a good defence in court. The next day Haywood wrote from Amherst, telling Roscoe he hoped his case would definitely determine the right of free speech in Saint John. "Jail birds are the birds of liberty," Haywood said.[4]

In police court that morning, Roscoe entered a plea of not guilty and went on the stand in his own defence. The report in the *Saint John Globe* said that Roscoe refused to take an oath on the Bible but agreed to affirm the truth by raising his right hand and offering to tell the truth. After hearing the evidence, the magistrate questioned Roscoe at some length. He then said that while some people look upon socialism with favour, others look upon it as a form of anarchy and disorder. "Supposing that what you said to those people last night excited them to revolution," the magistrate suggested. "It would be little less than anarchy."[5] He told Roscoe that he would have to stop holding open-air meetings in Saint John.

Roscoe replied that he always thought that as a citizen of Canada he could talk when and where he liked to. The magistrate, who seemed quite patient to Roscoe, said that if everyone was allowed to speak when and where they pleased, order would run riot. The magistrate asked Roscoe if he would promise to hold no more meetings on the public streets of the city. Roscoe hesitated and consulted one or two of the other socialists present. He said he didn't know what answer to give, that the right to speak in a public place was a matter of principle for him and that the magistrate was asking too much. After another short deliberation Roscoe promised to comply with the judge's request and was allowed to go.

Although he was heartily congratulated on his dismissal, Roscoe was troubled. He thought he should have forced the judge into either fining him or sending him to jail. A few days later, back home in Albert where he was working part-time at the nursery, Roscoe wrote a letter to *The Eastern Labor News*, praising labour leaders for their courage to go to jail in defiance of unjust laws. He closed by writing, "I am afraid I am not one of those heroic souls as my cowardice when hailed into your Saint John police court has disqualified me."[6] After Haywood saw the letter he wrote saying Roscoe should not be too critical of himself because other opportunities to show his support for socialism would surely be open to him.

Roscoe was impressed when he finally heard Haywood speak. "I found him a huge, likeable man and a natural orator. He could sway a crowd." Haywood was indeed a rough giant of a man who could inspire solidarity among workers. Broad-shouldered, his face pockmarked and scarred, he wore a patch over his right eye (which he had lost in a childhood accident). He dressed in a plain tie and dull suit. Anyone could tell from Haywood's appearance that he had suffered—working in the mines at the age of fifteen, battling police on the picket lines, and sitting month after month in the dampness of U.S. jail cells. Carrying his huge bulk to the speaker's platform, Haywood didn't need to resort to bellowing or bullying. He possessed genuine power and used simple language to rouse his audience.[7]

Haywood's celebrity status meant that his speeches were well attended and covered by the mainstream media. With great drama and bravado, Haywood described in detail the armed battles his union had waged to organize mines in the western United States. His message was that all workers had a common need to organize for political action, whether they were in the Maritimes or in Cripple Creek, Colorado. The depression that had started in 1907 was still hurting workers in the Maritimes, and Haywood said that men who could not find employment had a right to beg or steal, if necessary, before they starve. "The workers do not ask for charity," Haywood told a Moncton audience. "We despise charity. We want justice and we demand justice."[8]

Haywood got his biggest reception in Cape Breton, where he filled theatres several times with more than a thousand workers. He told the men who worked in the Nova Scotia mines about the health of workers in the lead mines, about how their joints became twisted, their hair and teeth fell out, and their nervous system became so damaged that they couldn't hold a glass of water to their lips without spilling half of it. Yet, he said, business took no responsibility for the workers' plight. After years of toil, the workers were left to survive as best they could in the mining town slums, the "capitalist hell." Haywood told the miners that they must achieve their own emancipation. They couldn't wait for some kindly saviour to do it for them. He talked about making the union an industrial school in which workers would educate themselves so that when the Socialist Party achieved political emancipation, the union would be ready to manage the mines.[9]

❧

At the same time that Haywood, Gribble, and Roscoe were preaching socialism to Maritime workers, a rival political movement appeared on the scene. Organized labour, frustrated with its unequal share in the proceeds of its own work, was establishing independent labour parties in several areas of the country. The workers behind this movement were mostly skilled men employed in manufacturing, construction, and mining jobs. Many of them had traditionally belonged to the Liberal Party and identified with the moderate socialism of the British Labour Party. But they found that these working-class politicians preferred to discuss issues of practical and immediate importance and often promised no more than "a square deal" for workers.[10]

"I am absolutely opposed to every form of compromise," Roscoe wrote in *The Western Clarion*. "I hold labor party politics worse than hell."[11] His strongly held views came from his conviction that labour political parties would do nothing to solve the problems of working-class people. He believed that a labour political movement would only serve to guarantee the continuation of the capitalist system because of the interrelated role of labour and capitalists. The one labour-oriented group that Roscoe did support was Bill Haywood's Industrial Workers of the World, which advocated replacing the capitalist system with a co-operative society.

Roscoe conceded that trade unions were a source of strength to the workers, but only as a weapon forged for the purpose of enabling the workers to make "collective bargains" with owners. "Their function...is that of bargaining and dickering over the terms or price of slavery. Never were they intended to abolish slavery nor could they do so even though they tried, with the limitations of their organization."[12] He said that the efforts of individual unions to increase the benefits of their own members merely created an "aristocracy of labour" among working people.[13]

In an open letter to unionists published in 1909 in *Cotton's Weekly*, Roscoe chided the union men for the gains they claimed to have made, such as larger wages and a shorter work week. He said their gains were illusory because as their wages increased their purchasing power decreased. "Can't you see that, even though you may, by a show of force, compel your employer to raise your wages, he, being the owner of the product of your labor, can increase the price of the commodities which you must have, and thus get it all back again?"[14] Roscoe also noted that Canadian unions were losing many more strikes than they were winning, quoting Department of Labour statistics to illustrate his point. He argued that labour had won fewer than 40 per cent of the more than five hundred major strikes between 1901 and 1907. Worse still, union workers had been replaced by scabs in 118 failed strikes.

When Roscoe and Gribble observed Labour Day celebrations in Moncton in 1909, they deplored the co-operation that existed between labour and business. Many companies used the parade as a method of promoting their products—the sort of collusion between labour and business that the socialists most detested. Dozens of small children gave out free Shamrock tobacco and British Navy chewing tobacco. Roscoe no doubt felt that many of these same companies paying tribute to labour would just as soon break a union or fire someone for questioning company profits. The labourers all tried to look sternly dignified. Roscoe said the whole thing reminded him of a "cross between an advertisement parade and a funeral procession."[15]

Although Roscoe and other party members ridiculed the labour movement, they still looked to the ranks of labour to recruit new members. Socialist Party members also sometimes supported union activities. Roscoe attempted to help factory workers in Amherst strengthen their union, and he supported the United Mine Workers in their battle against the coal company in Springhill in 1909-11. In Cape Breton several people were involved in both the Socialist Party and unions. In Saint John, key members of the party participated in labour activities, and one member of the local executive, Colin McKay, said he believed it was a mistake to sneer at the trade unions because such an attitude caused the unionists to remain aloof from the party. Later, both McKay and Fred Hyatt—whom Roscoe had met in

Calgary—became active in the Saint John Trades and Labour Council. Roscoe was opposed to such deep involvement by socialists in the labour movement, but he did not publicly criticize McKay and Hyatt for their involvement.

For their part, the craftworkers who dominated the labour political movement developed a deep distrust and dislike for socialists such as Roscoe because of the scorn heaped upon them. "In an age of relentless challenges from ruthless corporate employers," labour historian Craig Heron has written, "skilled workers placed defence of their workplace organizations above all else, and the ceaseless attacks from the left on these craft unions created a virtual unbridgeable gulf between socialism and labourism."[16]

In many Maritime communities labour unions were finding themselves under attack because a rapidly expanding industrial economy was becoming increasingly controlled by a small, interlocked group of finance capitalists based mainly in Montreal. The business interests lay more in short-term speculation than in long-term investment in industry. To attempt to protect their unions and their wages, Maritime unions—in contrast to the image of the docile Maritime worker—engaged in at least 324 strikes from 1901 to 1914.[17]

For the labour leaders who rejected the radical solutions of socialism, forming their own political organizations seemed like a logical step. During the period of 1908 to 1911 in the Maritimes, the Canadian Labor Party established branches in several centres, including Halifax, Moncton, Saint John, Pictou County, and Cumberland County. The new Labor Party was particularly well received in Halifax, where workers and the daily newspapers were pleased to see a fresh alternative to twenty-two years of Liberal government. John Joy, considered the "grand old man" of Halifax labour for organizing both the longshoremen and the street-railway workers, was nominated by the Halifax Labor Party to contest the provincial election. Roscoe tangled with Joy over the role of labour at a campaign meeting in Halifax.

The Nova Scotia election was held in June 1911, and while the socialists made no attempt to field a candidate in Halifax, Joy was soundly defeated despite the support of all three daily newspapers. Roscoe's fears that members of the Labor Party could readily be co-opted by the system seemed to be confirmed when the defeated labour candidates took jobs offered them by the Liberal government. John Joy was appointed to the Workmen's Compensation Board and active trade unionist Philip Ring was named the province's first factory inspector.

The one major labour centre in the Maritimes that did not succeed in maintaining a labour party was Cape Breton. The Socialist Party was too firmly entrenched in five locals in different communities on the island. The socialists ran their first candidate in the Maritimes in the 1911 provincial election.

❧

The labour political activity that interested Roscoe most was in Cumberland County, just across the Nova Scotia border not far from Albert. The main centres in Cumberland County were the booming industrial town of Amherst and the mining communities of Springhill and Joggins. Dissatisfied with traditional party politics, a collection of Amherst unionists and Springhill coal miners had got together in 1908 and formed the Cumberland Labor Party.[18] This occurred before the Socialist Party was well known in the county, and several members of the new Labor Party were more socialist than labour-oriented.

In May 1909, after the press had carried glowing reports of the Labor Party convention, Roscoe sent a letter to all three of the labour-oriented papers in the Maritimes urging the workers of Cumberland County to abandon the Labor Party and join the socialist movement. Roscoe's letter criticized the Labor Party's platform, which included a proposal for an old age pension at the age of fifty-five, a workmen's compensation act, legal recognition of labour unions, and "a minimum living wage, based on local conditions."[19] Roscoe said the minimum wage proposal was silly. "You propose to condone a system which makes the future of your children one of a miserable machine existence, provided—always provided—you can get dry bread and a bowl of soup for the present." He urged the workers to drop their demands for sops and insist on the absolute emancipation of their class from wage slavery.[20]

The socialists were working hard to make their party more attractive than the Labor Party to the workers of Cumberland County. Wilfred Gribble had spent a total of seven days in Amherst during his tour, addressing a dozen meetings. Socialist literature and copies of the two socialist newspapers were distributed among workers. The socialists established chapters in Amherst, Springhill, and Joggins. At the 1909 convention of the Cumberland Labor Party, delegates who favoured the Socialist Party introduced a motion that the Labor Party dissolve itself and declare for the Socialist Party. The resolution was referred to the membership and a referendum resulted in a two-to-one vote in favour of dissolution. But a convention held shortly thereafter decided to shelve the matter for a year. In the summer of 1909 a bitter strike was under way at the Springhill mines, interfering with the workers' political activities. In the meantime, the Labor Party nominated two candidates to contest the Nova Scotia election, which would come sometime in 1911.

Roscoe made his first working trip for the Socialist Party to Cumberland

County on the July first holiday weekend in 1909. The Albert baseball team was going by excursion boat across the Bay of Chignecto to play the miners' team at the village of Joggins, one of four small rural communities of miners who worked in the Joggins mines. Roscoe was able to obtain the use of the United Mine Workers hall without charge for the evening. About twenty-five people attended the meeting—not bad, Roscoe thought, considering there were baseball and football games on at the same time. Roscoe spoke for about fifty minutes, giving some of the history of the socialist movement and explaining why he thought people should support the Socialist Party. Toward the end of the meeting he signed up seven people for memberships in the Socialist Party. The workers also promised to sign up another twenty-five people within a few days and forward the names to him.

One of the men who stood out at the meeting was Jacob Resnick, a middle-aged Jewish merchant who had acquired his knowledge of socialism in Russia, where he said men and women were sent to Siberia if they were caught with socialist literature in their homes. Roscoe and Jake Resnick were to become life-long friends. Resnick pointed out that Roscoe showed too much impatience and aggression in his speeches about socialism, that he'd only anger people by calling them "wage slaves" or by saying that people who realized the benefits of socialism but were too afraid to join in the movement were cowards and traitors to their wives and children. Resnick advised Roscoe to change his ways and teach the rudiments of new ideas more cautiously, with humility. Resnick had a common saying, intoned with his Russian accent, that stuck with Roscoe for the rest of his life: "A man must creep before he can valk." Roscoe had to learn that he couldn't convince people by yelling at them and antagonizing them.

Roscoe was supposed to return to Albert on the excursion boat, but by the time the meeting was over the boat had left. He then decided to hold a street meeting so he could address the people who had gone to the sporting events. Later Roscoe estimated that during the evening most of the people of the little town had heard at least part of his oratory. This was Roscoe's first meeting with Nova Scotia miners, a group of men he would soon become well acquainted with and admire for their instinctive solidarity—but deplore for their short memories of the injustices they suffered.

❧

The next morning Roscoe decided to go on to Amherst, where a chapter of the Socialist Party had been set up several months earlier. In 1909 Amherst was a boomtown. After the construction of the transnational railway the town had rapidly become an important industrial centre. Between 1900 and

1906 its population nearly doubled, to ten thousand, as hundreds of people came in to work for companies making products as diverse as railway cars, steam engines, stoves, furnaces, shoes, woollen goods, luggage, and even caskets. At its peak, Amherst was in the top quarter of Canada's manufacturing towns.

Amherst looked well on the way to becoming the metropolis of the Maritime provinces. Proud of its bustling nature, town officials had placed a sign in bold letters at the end of the railway station platform: "Busy Amherst." Roscoe said that one night, amid the revelry and drinking that went on in the town, the sign was changed to "Boozy Amherst."

Many of the industries in Amherst were owned locally. Most of the town's businessmen had started their companies from scratch, become millionaires, and now lived on the town's exclusive Upper Victoria Street. The twenty most prominent capitalists served on one another's boards of directors, made financial contributions to the Conservative Party, controlled the town council, and held membership in the posh Marshlands Club.[21] Most of the working people in the town lived in rented houses and barely made enough money to support their families. Their interests tended to be simple ones, perhaps playing horseshoes, watching a ball game, or going fishing. If anyone in the family got sick and had to spend a few weeks in hospital, the family would have to scrape for a year or more to pay the bills.

Roscoe arrived in Amherst on a Saturday and that evening about a thousand people attended a hastily organized socialist rally on Victoria Square. Roscoe stood on the bandstand speaking for about an hour, and later that evening about one hundred people interested in the Socialist Party attended a meeting at a local hall. The next day the *Amherst Daily News* commented that this was the first time many people in the town had ever heard the doctrines of socialism presented. Concerned about the inequities it saw in the community, the paper carried an editorial criticizing the men who controlled Amherst. "There were many homes in Amherst last winter where children had not suitable clothing for a Nova Scotia winter and where, on more than one occasion, they had to go to bed with their cravings for food still unsatisfied," the editorial stated. "It was not because the breadwinner of the family was a drunkard or that he was shiftless and dishonest. It was not because there was not enough to spare in the land, it was simply because he had no work and could find none."[22]

The newspaper said that half a century before there were no great industrial corporations in Canada. At that time, it said, production was carried out by individuals for local use, and the richest was poor and the poorest was rich.

> Conditions today have entirely changed. The wage earner has little or no control over the product of his labor. Instead of working on his own account in his own small way, he toils in large factories and other undertakings under employers who own and control the capital embarked in them. Labor has ceased to be individual. It is now collective and men who are only receiving a living wage are ever haunted by the spectre of 'out of work.' What the end will be we do not pretend to know but this fact is potent to all—that the big question in our Canadian Life is not the building of Dreadnoughts and invincibility, it is not the unification of the Empire, but it is the working out of the problem that will ensure steady employment to the wage earners of Canada.

Soon Roscoe was able to spend more time organizing for the Socialist Party in Cumberland County because he got a job working for W.A. Fillmore's nursery, located near Amherst. W.A. Fillmore, not directly related to Roscoe, had once been a partner in the Albert Nursery. Roscoe was probably the only young person in the Maritimes with considerable nursery experience, and he took charge of the planting operations at the nursery. In his spare time he continued to do what he could for the cause of socialism. W.A. had known Roscoe for several years and was fond of the young man—even though he told his employees to pay no attention to the lad's political ranting.

Roscoe became close friends with several Amherst men who were interested in radical politics and became members of the Socialist Party.[23] Most of them took part in trade union activities, and although Roscoe detested labour unions he helped with labour organizing in Amherst. He got involved in one dispute involving the town's black workers. About five hundred blacks lived in Amherst, and even though many of them worked in the big companies, often in the most difficult jobs, they had not been invited to take part in labour activities. Blacks were barred from Amherst hotels and many restaurants. When there was talk of formally barring them from the Car Workers Union, Roscoe and a handful of socialists strongly objected. Roscoe refused to speak at a meeting unless the colour bar was dropped. With the support of the union executive, the black workers were admitted.

During this period the economy in Amherst began to change for the worse. North America had been hit by a recession and industrial growth in Amherst slowed. Throughout 1908 factories had reduced wages and laid off workers, and some of them had suspended production. The national economy gradually recovered, but Amherst would never be the same. After the recession capital investment was concentrated in central Canada, and central Canadian companies used their competitive advantage to force Amherst firms into bankruptcy.[24] In other cases companies used industrial mergers to gain control of other large firms. The Rhodes-Curry Company, which had manufactured thousands of railway cars for the great Canadian railway expansion, merged with another central Canadian firm, reducing Amherst's most important industry to a division within a large corporate structure based outside the region. In the years following the integration, the new Canadian Car and Foundry Company concentrated on modernizing and expanding its central Canadian operations while simultaneously reducing production in Amherst. The Amherst plant never completed a conversion from wooden railway cars to steel-pressed cars. In general, the infusion of investment capital into Amherst dried up, and a town that employed four thousand people in 1907 would employ only seven hundred people by the mid-1920s.

Roscoe and the other socialists, not surprisingly, attacked this company practice of shifting huge amounts of investment out of one part of the country to pursue greater profits elsewhere while at the same time causing the town of Amherst and its citizens to suffer. The Amherst workers also tried to fight back to keep their jobs and their level of pay, but they weren't always successful. In 1910 Roscoe was in Amherst to witness one such struggle by workers at the Canadian Car and Foundry Company. Most of the twelve hundred men there were pieceworkers and in the practice of staying on the job only until they had earned what they felt was a good day's pay. But a new plant manager soon discovered that the men were making a good wage by working for as little as seven or nine hours a day. Soon about 150 men working in one division of the plant noticed a change in their piecework payment that amounted to a 12 per cent wage cut. When the company made a second cut the men held a meeting and agreed to walk off the job. Roscoe was told they were earning less than a dollar a day.

At first the men talked boldly of holding out for weeks. The company said it would make no concessions. Roscoe and his socialist friends called a meeting, advertising the event in the local paper. Roscoe hoped the meeting would lead to the formation of a stronger union for the workers and also allow him to promote the Socialist Party. "Fancy our surprise when we reached the hall to find only a half dozen on hand and not one of them was a striker," Roscoe wrote. He learned that "a number of the loudest talkers" among the strikers went to the car works and worked as scabs. "Some of these were ex-members of the SP of C, and have always been foremost in telling what they proposed doing 'a little later' in reorganizing Amherst, carrying on propaganda work, etc."[25] The next day the men went back to work. The company convinced them that the readjustment had not cost them a loss of pay.

Roscoe was disappointed by what he saw as a lack of courage on the part of the workers, but he didn't believe the chances of the Socialist Party in Cumberland County were seriously damaged. Workers from Springhill and Joggins as well as Amherst were key to the future of the party in the area. With a solid base of about forty members in Amherst, Roscoe felt sure that when the miners' strike at Springhill ended, the workers would unite behind the Socialist Party to challenge the old-line parties. What would happen in Springhill during the long strike of 1909-11 would be instrumental in determining the future of the Socialist Party in the Maritimes.

SIX

Class Struggle in Springhill 1909–11

THE TOWN OF Springhill is situated on a wind-swept hill in the middle of Cumberland County, Nova Scotia. Deep beneath the barren, rocky landscape ran five rich seams of coal. In the 1870s, investors from Halifax and Saint John had sunk the first large commercial mine on the hill, and men and boys were taught crude methods of digging the coal and loading it onto the railway cars that took it to be sold in the large towns and cities of the Maritimes.

The men who worked the mines of Springhill quickly won a reputation for being the most militant work force in the Maritime provinces.[1] As early as 1879, when the town was becoming the largest single producer of coal in Canada, they showed their determination by forming the first miners' union in Nova Scotia and going on strike in response to a wage cut. During one twenty-year stretch the miners went on strike twenty-seven times.[2]

By 1907 the struggle against the Cumberland Railway and Coal Company was going badly for the Springhill workers. Their union, the Provincial Workmen's Association (PWA), had begun a series of conciliation board hearings, which had failed to resolve their differences with the company. It seemed to many of the workers that the PWA leaders had been co-opted by the company, and they began to consider joining the U.S.-based United Mine Workers of America (UMW), in hopes that a large union would be better able than the regional PWA to stand up against the coal companies.

In a 1908 referendum, only 53 per cent of the coal miners in the province had voted to join the UMW, but the Springhill miners demonstrated their militancy: Cumberland County miners voted 90 per cent in favour of the

UMW. The Springhill miners established Local 469 of the United Mine Workers of America, the first local in the province. When the coal company refused to recognize the new union over the less powerful PWA, the Springhill miners and other miners in Cape Breton were prepared for a major confrontation with the coal companies over—among other issues—recognition of the UMW.

Roscoe Fillmore and other socialists were particularly interested in the Springhill miners because of their long record of labour militancy. Roscoe thought that the miners, stymied by the powerful coal company, would be receptive to the socialist movement, and he hoped they could be convinced to abandon the Cumberland Labor Party and join the Socialist Party of Canada. When the party made its first foray into Springhill in May 1909, organizer Wilfred Gribble was bowled over by the reception. "I think Springhill must have broken the record in number of names on a charter application," he wrote. "This is a mining camp and is seething with revolt." Gribble, who had organized several Socialist Party chapters in Ontario, reported that Springhill was "the ripest place for organization" he had yet found.[3] With thirty miners taking out party memberships, Springhill became Nova Scotia Local No. 2 of the Socialist Party of Canada.

Some of the miners were especially receptive to the ideas of radical socialism after almost three years of frustration with conciliation board hearings. The biggest blow came in July 1909 when the conciliation board, which had been sitting over a two-year-period, ruled against the PWA on all main issues.

In Cape Breton especially, UMW miners found themselves in a difficult situation. The companies had harassed UMW leaders and by April 1909 had fired all of the top UMW spokesman as well as more than a thousand men who had joined the union. To work the mines they imported more than two thousand loyal PWA members and several hundred out-of-work transients from Halifax and Montreal. In addition the companies hired more than six hundred goons and armed them with guns and clubs. Even though UMW leaders didn't see how they could win under such difficult circumstances, union members declared a strike on July 6.

The sixteen hundred Springhill miners, almost totally united under the UMW, felt they were in a much stronger position than their counterparts in Cape Breton. They went on strike on August 10. The company, claiming it was losing money, reacted to the strike by posting notices throughout the town saying the miners could return to work under the old terms. According to the company, if the workers refused the mines would not resume operations unless the men accepted a 10 per cent reduction in pay.

❧

When Roscoe visited Springhill during the summer of 1909, he found a town full of miners braced for a long strike. The Cumberland Railway and Coal Company was the only major employer in Springhill, but it didn't control the town in the way other companies controlled one-industry towns. The Springhill miners had established a certain independence: They believed that they, and not the company, controlled the functions in the mines and that they had the ultimate power to close down the mine.[4] Most of the 5,800 people who lived in the town were solidly behind the miners. While miners in other places bought all their goods from company stores, the Springhill miners purchased their supplies from downtown merchants. Main Street, the central commercial area, had more than a hundred stores and businesses along with three hotels and a weekly newspaper. Storekeepers and property owners reduced prices and rents until the strike was over. Many of the miners owned their own homes, again in contrast to the situation in other mining towns, where workers tended to live in company-owned housing. In addition, the UMW gave out strike pay: A married man got two dollars a week for himself, a dollar for his wife, and fifty cents for each child, much less than the three dollars a day or so that many of the men received for their work in the mines. The UMW also bought woodland so families would have a free source of fuel.

Springhill's strong community spirit was symbolized by a sixteen-foot-high statue, the White Miner, standing on the highest point of Main Street. The grey marble monument was built to honour the 121 men and boys killed in a terrible, sudden "bump" in 1891. The statue depicted a miner ready for work with a pick in one hand and a safety lamp hanging from his belt and was a reminder of the constant danger of the work. In the years following the disaster of 1891, another forty men died in the mines—an average of one every four months.[5] Few families had not lost a loved one to the mines.

Working under such conditions led to a strong solidarity and camaraderie among the miners. When the strike began the *Amherst Daily News* stated that Springhill had always been a union town: "The name of 'scab,' or as it is locally termed 'blackleg,' means disgrace and the enmity of the [union] lodge for years to come."[6] The miners believed they could hold out for several months.

Roscoe saw the potential of the strike as a radicalizing process for the miners. He hoped that their conflict with capitalism combined with the strike's promise of bitter experience would push them toward the socialist movement. Of course, a lot depended on conditions largely outside his and the other socialists' control. What tactics would the company use? How would the strike develop, and how long would it last? How many miners and their families would see socialism as the answer to the problems they were facing?

It seemed likely to Roscoe that the miners would grow to hate a system that made them risk their lives daily to earn a small income while the general manager of the company, J.R. Cowans, risked nothing and lived in luxury. Cowans's huge house, which the boss liked to refer to as the Mansion, dwarfed other homes in the town. His family had a staff of at least six maids. They had a huge summer home at Parrsboro. Cowans, a horse lover, owned three thoroughbreds and paid two trainers to look after them. Worst of all, Roscoe disliked Cowans's attitude toward the miners and the community. One long-time resident remembered, "In church on Sunday evening when Mr. Cowans bowed to certain people this meant that they were to go to his home for after-church supper, and since it was like a royal command, no one refused."[7]

Several leaders of the UMW strike committee were socialists, including William Watkins, the UMW secretary and main spokesman for the miners, and Seaman Terris, a respected union official who had represented the miners on one of the conciliation boards. Two other socialists, Calvin Ward and D.C. Matheson, played a role in the strike and became members of Springhill town council during the strike. *The Halifax Herald* referred to this group of men as "the socialist ring of the UMW."[8] Many other rank-and-file miners became interested in the party during the strike. In addition, some of the organizers who came to Springhill from UMW headquarters in Pennsylvania were also socialists.

The most energetic booster of socialism in Springhill was a Belgian-born man. Jules Lavenne, who spoke no English when he immigrated to Nova Scotia, had lost a leg in an industrial accident and was given work in the machine shop of the coal company. He had preached the doctrine of socialism in the town until the Socialist Party began to organize, and he became one of its first members. Soon after the strike began, Lavenne won a competition by selling more subscriptions to *Cotton's Weekly* than anyone else in the country.

Roscoe's job as foreman at W.A. Fillmore's nursery in Amherst kept him close by during the strike, and he was able to visit Springhill on a regular basis. In Springhill he usually stayed at the home of Socialist Party member Seaman Terris. Roscoe's main role was to serve as an agitator for the Socialist Party. He spoke at socialist rallies and union meetings, urging the workers to fight for their rights, and spent time on the picket lines. Roscoe also became a reporter during the strike, covering events for *Cotton's Weekly* with colourful, opinionated prose. Throughout the strike socialists such as H. H. Stuart of Newcastle and Sophie Mushkat of Moncton visited Springhill to help Roscoe, Lavenne, and the others.

Roscoe was deeply stirred by the drama of the strike. In a report for *Cotton's Weekly* he described the scene one morning when he and Seaman Terris joined the protesters on the picket line near the mines, which were bunched together at the south end of Main Street.

> Although it was only a few minutes after six when we arrived, quite a number had already gathered and were patrolling the streets. About seventy in all were on hand for the work. About twenty-five of the fifty or sixty thugs and gun men in the employ of the company were on hand for the purpose of aweing the men. These thugs were heavily armed, being provided with Colt .44 calibre revolvers, and several of them carried clubs. Some of them were so bold as to carry the revolvers well in sight, one fellow I particularly noticed carried his gun in his vest pocket.[9]

During the following weeks Roscoe was surprised by how many people on the picket line and in the strike wore red buttons, signifying that they were socialist supporters. "Probably eight hundred are worn now," he wrote, "and from talking with the bearers, I have concluded that the most of them know mighty well what the red means." He was encouraged by the number of women taking a turn on the picket line. "Many of them are as well grounded in socialism as the men and just as fearless in propagating it."[10]

The fact that Roscoe supported the Springhill miners and the UMW in their fight against the company did not mean he had changed his opinion about unions. He still believed that society was in need of a radically new structure and that unions were misguided in their approach to solving problems through negotiation with capitalists. Many of his fears about the power of corporations, he thought, would become evident during the Springhill strike.

Early in the strike the UMW charged that the two main companies, Cumberland Railway and Coal and Dominion Coal, were in collusion to fix coal prices and establish a monopoly in the Maritime coal industry. The close relationship between the two companies was evident as Dominion Coal recruited scab workers for Cumberland Railway and Coal. The union forced the companies into court, where considerable evidence of collusion was brought forth. The companies had developed a strategy for setting wage rates and had an agreement to keep coal prices artificially high, particularly within the Maritime provinces. Unfortunately, the law regarding combines was so weak that the companies were found not guilty.[11]

Another problem was that the companies were not required by Nova

Scotia law to recognize a union chosen by the workers. J.B. McLachlan, the Cape Breton mineworkers' organizer, attacked the arrogance of the companies:

> The man who has the pluck to stand up among his fellows and advocate any trade union not approved by his employer shall do so at the peril of having sentence of death by starvation passed upon him and his family. Our 'captains of industry' who never tire in their hypocritical ravings about 'individual initiative' demand that their every employee shall be docile, obedient and tractable to all their wishes, especially in regard to trade unions.[12]

❧

Three months into the strike, a by-election was held in Cumberland County to fill a vacancy in the provincial legislature. The socialists were interested to see how their Labor Party rivals would do at the polls. The Labor candidate, Adolph Landry, had worked in the mines during the terrible explosion of 1891 and was a university graduate and manager of the Amherst correspondence school. The Liberal candidate, E.B Paul, was a staunch UMW man and had been a fixture of the Springhill labour movement since the 1880s. His family had been one of the first mining families in the town, and his party had been in power provincially for twenty-two years. *The Halifax Herald*, a Conservative paper, went so far as to suggest that the Cumberland Tories should endorse Landry instead of running its own candidate, so the Labor candidate would have a chance of winning a seat from the Liberal government. In the end the Conservatives did enter a candidate.

Landry was a good speaker and ran a strong campaign, visiting small outlying communities and drawing more than a thousand people to a rally in Springhill in the closing days of the campaign. On an election day marred by a severe storm, however, he came a distant third to Paul—though his total of 1,278 votes was still more than double the total that old-line party strategists predicted for him. He received more votes in the town of Springhill than his two opponents combined, and he got a large vote in Amherst, but he did poorly in rural areas. Roscoe thought that if a Labor or socialist candidate were ever going to be elected in Cumberland, he would have to do well throughout the county.

Four months into the Springhill strike, the Cumberland Railway and Coal Company began to take steps to get the mines into production. In December it started bringing scab workers in from other parts of Canada. But, as Roscoe recorded, when new workers arrived, an equal number quit the mines. Although the company boasted that it raised 3,500 tons of coal during February, Roscoe figured the true figure to be about half that. Roscoe said that in March 1910 only about a hundred men were working, and many of them were inexperienced miners.

Later the company began recruiting foreign miners, mostly German and Belgian. With the active help of the Canadian government the company advertised abroad that excellent work was available in Canada. The Europeans usually arrived at Halifax. "In most cases the company, with its armies of police as well as the military at its disposal, was able to bring the men from Halifax docks without the miners being able to contact them," Roscoe wrote years later in his autobiography. "A high fence surrounded the mining property and living quarters were provided inside the fence or 'bull pen' as the strikers called it. But the newcomers hadn't been long in the bull pen before the news leaked to them that they were scabbing." The German and Belgian miners were mostly socialists, according to Roscoe. "The first breaks were made when strikers lined up near the bull pen fence and sang socialist songs. Recognizing the tunes, the men inside the pen came out and practically all of them joined the strikers." Roscoe was able to interview a number of the German miners and discovered that many of them were members of the German Social Democratic Party and the country's miners' federation.

During the winter of 1909-10 Roscoe and the socialists were doing everything they could to convert the miners to socialism. They frequently spoke at meetings and sometimes held their own rallies. In February more than nine hundred people crammed into the Opera House in Springhill to attend a Socialist Party rally and hear speeches by Roscoe, Seaman Terris, and Sophie Mushkat. Roscoe talked about the class struggle for an hour and fifteen minutes and had to stop speaking several times because of the cheers and applause.

The church was an important part of the lives of most miners and their families, and the socialists tried their best to convince townspeople that their party wasn't against religion. Still, some of the socialists couldn't restrain themselves from lashing out at the ways of the church. In particular they talked about how the Presbyterian Church catered to general manager Cowans and came close to being the "company church." Cowans would enter the church with a flourish, and the service would not begin before he had settled in his pew. Many Springhillers later recalled the large donations Cowans made to the church and the beautiful silk dresses worn by his wife.[13] In one sermon an Anglican minister said that the cause of labour was just but warned his congregation about the socialist element. He said they should not listen to the revolutionary socialists who preached materialism and class warfare. "The rights of man have been thundered in our ears by itinerant agitators of late," he said, "but we hear very little about the rights of Him who is said to have created this world and all that are therein. HAS HE NO RIGHTS?"[14]

H.H. Stuart, who was a lay minister as well as a socialist, came to Springhill especially for the purpose of giving lectures on how socialism and religion could co-exist. Terris wrote in *Cotton's Weekly* that he knew some people incorrectly had it in their minds that socialism would do away with religion. "I am glad to report that I see no reason why I should leave my church or drop my religious views by trying to improve my conditions materially. We have one local preacher in our [socialist] local and many members of quite different religious beliefs, and I hear no kick about religion."[15] Still, many miners remained convinced that the socialists wanted to do away with the church.

The socialists' opportunity to take over the Cumberland County Labor Party finally came in May 1910 when the party met in Maccan, near Amherst, to discuss its future and reconsider the resolution introduced at the 1909 convention, where members had voted by a two-to-one margin to endorse the platform and principles of the Socialist Party.

The Maccan meeting was chaired by Roscoe's friend Dan McDonald of Amherst, who was sympathetic to the socialists. Roscoe, who was allowed to be present as a guest of the Labor Party, waited with anticipation as the meeting discussed whether the Labor Party should be dissolved. Seaman Terris, strongly in favour of joining the socialists, stated: "This ILP [Independent Labor Party] bears the same relation to the Socialist Party as does the old sectional PWA to the international UMW. It is obsolete."[16] When a vote was held, the socialists won. The Cumberland Labor Party was dissolved, and its funds and property were turned over to the Socialist Party.

A meeting of the Socialist Party was convened immediately. Roscoe was elected chairman and William Watkins, the UMW secretary, was elected secretary. The people present who had not previously joined did so, and a few other socialists came into the hall and joined the meeting. Terris and Adolph Landry of Amherst were chosen as the Socialist Party candidates for the next election. Terris accepted the nomination, but Landry was not at the meeting. Roscoe told the meeting that Landry would also accept the nomination even though Roscoe had doubts about whether Landry was really a socialist and whether he would join the new party. If Landry did not run, Roscoe himself was willing to be a candidate. Pleased with what had happened, Roscoe wrote soon after, "When the strike is over and the men recover somewhat financially, you'll see the red bloom."[17]

The takeover of the Cumberland Labor Party was front-page news in *The Halifax Herald* and led to a flurry of socialist activity in Springhill, where the

Socialist Party quickly signed up ten new members. The party made plans to hold street meetings in the town, to send speakers into the countryside, and to meet with workers employed at sawmill camps. But even though Roscoe was enthusiastic about the reception he and the other socialists were receiving, it was still unlikely that anywhere near half the miners supported or endorsed the politics of the Socialist Party.

❧

In Cape Breton the UMW miners on strike were losing their battle against the companies and the PWA strikebreakers. The company was aided by the presence of the militia, imported scab workers, and court injunctions. The strike was brutal. Families forced to live in tents saw their children die of the cold, disease, and malnutrition. In April 1910 the press published rumours that the company was going to greatly increase the number of scabs. Afraid they would lose their jobs permanently, many of the men returned to work. Their strike ended after nine months.

In Springhill the company was unable to get the mine back to a reasonable level of production because the strikers convinced many of the imported scab workers to abandon their jobs. The company decided it would have to adopt new, tough tactics. It started by evicting the thirty families that lived in company-owned homes. It also sought a court injunction to make picketing illegal on the grounds that the picketers were damaging the company's business.

The court immediately issued an injunction restraining the United Mine Workers from talking to or in any way interfering with the employees of the company. Most of the strikers ignored the court order, and more strikebreakers were convinced to leave their jobs and join the picket line. Soon nineteen miners—at least four of them socialists—were served with a restraining order. In spite of the injunctions the men continued to picket the colliery and, in fact, became more active than before.

Exasperated, the company decided to take a test case to court. Jules Lavenne, the Belgian miner and Socialist Party member who had been instrumental in persuading about forty German miners to quit the company, was charged with contempt of court for failing to obey the court order restricting the miners from harassing the company's business. Lavenne was indignant. "I did not obey the injunction because no injunction was issued restraining the company from bringing in scabs," Lavenne said in an interview. "If it is proper for the company to fetch in scabs it is just as proper for me, acting in my economic interests as a worker, to endeavour to keep them out."[18]

Cotton's Weekly, assigning Roscoe to cover developments in Springhill, said that Roscoe would report "from the class conscious revolutionary standpoint."[19] The newspaper started a special fund to pay for Roscoe's expenses, and donations came in from miners' unions as far away as Cobalt, Ontario, and Greenwood, British Columbia. Roscoe's bias against the courts and judges was clear: They were both used to enforce laws that favoured and entrenched many of the negative values of capitalism.

While the company was allowed to make full use of the law to undermine the effectiveness of the strike, it conveniently ignored a Nova Scotia mining regulation stating that before a man could work as a miner in the province he must be certified by a board of examiners and have worked in some capacity in the mine for at least one year. None of the imported scab miners employed by the company met these criteria. At the urging of the union, at least three men who had worked as scabs signed affidavits declaring that they did not have a Nova Scotia certificate to be a miner. The government didn't take action against the company.[20]

Lavenne was ordered to appear before a Halifax court in June. Roscoe accompanied Lavenne on the train trip from Springhill to Halifax, and they soon noticed they were being followed by two men who seemed to be company spies. That night Roscoe and Lavenne stopped to stay over at a hotel in Truro, halfway between Springhill and Halifax. Surprisingly, just when they were getting ready to go to bed the proprietor came to their room, accompanied by one of the men they had seen on the train. The proprietor asked Roscoe and Lavenne to wake up the strange man and his friend in the morning in time to catch the six o'clock train to Halifax. The next morning Roscoe and Lavenne woke the men up as requested and went for a walk through the town. When they spotted one of the men lagging along behind they confronted him, but he wouldn't admit that he was tailing them.

"We got on the 6 a.m. train and the spy followed suit," Roscoe reported in *Cotton's Weekly*. "When the train pulled out, we remained on board until she was running fast and then jumped, leaving our friend to go on ahead and prepare the way. We took the next train, an hour later and were just congratulating ourselves when our spy boarded the train at Shubenacadie with another spy. From there on we had three of them watching us until we arrived in Halifax."[21]

In Halifax two of the men followed Roscoe and Lavenne to the court house. Roscoe speculated in *Cotton's Weekly* that the reason the company had them followed was to keep tabs on Lavenne so he wouldn't flee the area to avoid the possibility of going to jail. If this indeed was the reason, the company's efforts were in vain. Despite Roscoe's bitter feelings about judges, this one dismissed its attempt to imprison a worker. The company also tried to get Lavenne convicted of "wilful and corrupt" perjury, but that failed as well. If anything, the company's actions boosted the morale of the workers and the Socialist Party. Despite Lavenne's acquittal, the injunction against picketing remained in effect.

Throughout the strike, private police—like the men who had followed Roscoe and Lavenne—tried to provoke the workers and incite them to violence. One time Roscoe saw one of the Thiel detectives who had started causing trouble being beaten by a group of strikers. One of the detectives fired his revolver and was then passed over to the town police. He was charged with discharging a firearm and fined ten dollars. There were also altercations among the women in the town. Once the wives of two strike-breakers left the company's domain in search of a dressmaker. According to Roscoe's report, "They walked around the town for some time and finally got into wordy warfare with several of the town women—the wives of strikers. A battle ensued in which the strangers were decisively beaten—one being, it is claimed, rendered unconscious." The police arrived and arrested eight women. During the trial seven of the women were acquitted. In the eighth case the court imposed a fine and costs of twenty dollars. Roscoe, infuriated, said that the only evidence was given by a feeble-minded girl of about thirteen. "In this way the Springhill police court again put itself on record as a link in the chain with which the workers are shackled," he wrote. "Not one iota of real evidence was secured against the accused Mrs. Lounsberry, but the court, seemingly thinking that something must be done to awe the militant women, imposed a fine."[22]

The miners also had to deal with a sensationalist mainstream press outside of Springhill, which issued warnings about the dangers of anarchy in the small mining town. In response the UMW's William Watkins quoted from Springhill court records to show that between December 1909 and July 1910 only six miners and citizens were convicted of unlawful acts such as assault and disturbing the peace, while thirteen company employees were convicted of such crimes as discharging firearms, carrying a sheath-knife, and using insulting language.[23]

Nonetheless, the press reports helped give the company an opportunity to make an important strategic move. Arguing that it needed protection for its property and workers, the company asked Halifax to send in the militia. Watkins believed that the real reason the company wanted troops was so it could reassure a scab work force.[24] The provincial government agreed to provide militia members. Mayor E.A. Potter refused to sign the granting order, but the company got the necessary authorization from the county judge.

A contingent of 170 armed militiamen from Halifax arrived in Springhill in June 1910 and set up a military camp on company land about halfway between the town and the mine. A machine gun was ominously prominent in their arsenal. Roscoe had predicted that the military might be used to break the strike. Miners had lost several previous coal strikes in Nova Scotia after the intervention of the militia.

Roscoe wrote that the soldiers were sent to Springhill to frighten the strikers and other citizens, but there was little for them to do. "They have, at the time of writing, been here for almost four weeks and they have not yet done any police duty. Their most arduous work has been doing the usual sentry duty incidental to any military camp and playing football."[25] Sometimes the soldiers marched in formation out of town a couple of miles and then back. With their presence Roscoe noted a remarkable change in the attitude of the company thugs and scabs; now the hired guns walked boldly through the town and appeared to be looking for trouble. The militia did help to protect the strikebreakers and break up crowds of men and women in the streets of the town.

In June, when the last thirty families of striking miners who lived in company-owned houses were given eviction notices, Roscoe spoke to workers through *Cotton's Weekly*: "How do you like it, you fellows who have the callous lumps on your palms from having built the houses, dug the coal, built the railways and, in short, performed every task that has made the earth a better place to live on? How do you enjoy seeing the wives and children of your fellow workers kicked out on the street to suffer and starve?"[26]

August 10 marked the first anniversary of the strike, and the miners and their families held a big celebration. The main event was a giant picnic held on a farm outside of town and attended by more than four thousand people from several surrounding communities. In the morning about four hundred men came down Victoria Street and through the town in a parade led by the town's junior band. The socialist contingent of about a hundred men was led by Roscoe. As they marched along, a large red flag fluttered in the breeze and many houses along the route displayed red flags. "Red, red, red was everywhere," Roscoe wrote a few days later. "Red buttons and flags, rosettes, and hair bows. Many of the horses bore red ribbons."[27]

Another section of the parade, led by Jules Lavenne, consisted of a number of children who carried small red socialist flags. Lavenne had organized the children based on socialist ideals, calling them the Socialist Young Guard. The Young Guard, which Lavenne saw as the socialist alternative to the Boy Scouts, had its own code of ethics. The children were urged to love learning and be grateful to their teachers, to always stand up for their rights, and to resist oppression.

Many of the boys in the parade worked in the mines—in fact close to three hundred boys worked at the Springhill mines. While the employment of children was an issue in some areas of Nova Scotia, many coal-mining families reluctantly allowed their boys to work in the mines because they needed the extra income. Boys of twelve or fourteen earned about half the pay of an adult labourer, and most of them welcomed the opportunity to work in the mine because it meant initiation into the world of manhood. Their main jobs were underground, work such as opening doors that controlled the ventilation inside the mines so horses could carry out loads of coal. The boys were considered to be apprentices; once they had acquired certain skills and were older, they could go on to higher-paying jobs. When the 1909 strike began, about a thousand boys were working in the Nova Scotia mines.

New provincial legislation that came into effect during the Springhill strike prohibited any boy between twelve and sixteen who had not completed grade seven from working in a mine. However, boys who needed the income continued to do the work, and teenage boys would be a regular part of mine life and work for many years to come. Neither the union nor the socialists concerned themselves with the question of whether boys should be subjected to such hazardous working conditions or whether, at least, they should have an adequate education before they entered the mines. For its part, the union was more concerned with preserving its own job classifications.[28]

In December 1910, as the strike entered its eighteenth month, the coal company resorted to one of the great strategies of business in a time of adversity. The controlling interest in Cumberland Railway and Coal Company was sold to the Dominion Coal Company. The merger created a much larger and more powerful company. Dominion Coal, owned by wealthy business interests in Boston and Montreal, already had a privileged existence in Nova Scotia. In 1892 it had been set up on the initiative of the Nova Scotia government with a ninety-nine-year lease on all the unassigned coal resources of Cape Breton in return for a minimum annual royalty of 12.5 cents per ton of coal mined. Now, with the merger of the Cumberland company, the company controlled most of the coal fields in the province. The company's Cape Breton mines were in full operation, bringing in considerable revenue from the sale of coal in both Canada and the United States.

Although the miners didn't seem to realize it, the sale of the company to a firm that had the resources to continue the strike indefinitely was a decisive turning point. Roscoe had predicted all along that the Cape Breton miners would lose their strike, but he had thought the Springhill men had a chance. Now he realized the picture looked bleak.

Soon after the takeover, the company tried to further intimidate the strikers. A group of strikers who needed fuel to keep their homes warm went to a site outside of town and gathered a few bags of coal. The company sent its private constables and arrested six men, charging them with stealing ten dollars' worth of coal. The men were convicted in court and—given the option of paying a fine of five dollars and costs or spending thirty days in jail—they opted for jail. Another seventeen men were charged with theft of coal when they tried to retrieve discarded pieces of coal at the company's dump.[29]

Aware of its new strength, the company offered to settle the strike, but on its terms. In February 1911 it told the men that the strike could be settled if they would agree to quit the UMW, take a reduction in pay, and accept that only one hundred of the striking men would start work if the strike were ended. The men rejected the offer, but during the next few weeks any chance of winning the strike began to disappear. The rival PWA was reorganized in Springhill, and some of the UMW workers began to return to work.

❧

In May the strikers received the final blow. The UMW had several thousand workers out on strike in the United States and, to the surprise of the Springhill miners, it announced that because of financial difficulties it would no longer be able to provide them with strike pay. The union had already provided a huge sum—about $700,000—in strike pay for the Nova Scotia miners. Despite the opposition of many of the miners, an agreement was reached on May 27, 1911, to end the strike. Springhill's strike had gone on for twenty-two months. It was the longest strike in the industrial history of Nova Scotia and one of the most bitter strikes ever in the country.

The workers were soundly defeated. The United Mine Workers would not be recognized, for the present, in Springhill or in the rest of Nova Scotia. The PWA was soon back in control. For individual miners, the greatest hardship was the fact that when they returned to work many of them would be paid 10 per cent less than they had been paid two years previously. They were given some minor concessions, such as a fairer system of docking pay when rock was loaded along with the coal, and a new more agreeable manager had replaced J.R. Cowans.

As for the town of Springhill, it would never be the same again. Until 1911 no other area could match Springhill for its contribution to working-class protest. But defeat in the strike was savage and total for the miners and the community. Thereafter a profound helplessness replaced the miners' traditional activism.[30]

The defeat of the miners was also a serious blow to the Socialist Party, which was hoping to run a candidate in the June 1911 provincial election. Neither the party nor the miners had money to spare, and morale among their miner supporters was also understandably low. Roscoe and Jules Lavenne had gone to Cape Breton separately to ask the Maritime executive of the party for support, but they received little money. With only a few weeks to go before the election, Lavenne headed a campaign to raise $600 to contest it. It would cost $200 alone to register a candidate. An appeal for funds that went out across the country brought in small donations. To build interest in the socialist campaign, popular Alberta socialist MLA Charlie O'Brien addressed eight meetings in Cumberland County during the last few weeks before the election. But in the days leading up to nomination day, the party still hadn't raised enough money. The *Amherst Daily News* chided the socialists, saying that Roscoe, "the man who was to take a hand in the battle of the masses against the classes," had not been heard from for some time.[31] In the end the party did not field a candidate. Roscoe, disappointed and bitter that big business again had defeated the workers, finished his work at the nursery in Amherst and returned to Albert.

The Socialist Party was unable to maintain the position it had built for itself in Cumberland County. The fragile party structure crumbled. In December 1911 the charter of the Springhill local was revoked by the Maritime executive for non-payment of dues. Individual socialists in town continued to show an interest in the party, but with both morale and financial wherewithal at a low ebb, organized socialism in Springhill disappeared. The party chapter was never reinstated.

SEVEN

"Yours in Revolt, Roscoe Fillmore"

THE SOCIALIST PARTY of Canada was unable to maintain the network it had built up in the Maritimes. Roscoe Fillmore's little Albert chapter collapsed, and the building on Main Street known as the Socialist Headquarters became a tailor shop. By 1911, of the fifteen original Maritime chapters, only four—Saint John, Amherst, Glace Bay, and Sydney Mines—were still in operation.

Only two chapters—Glace Bay and Saint John—had any consistent level of activity. The busiest of the remaining socialist centres was Cape Breton, where the party entered a candidate in the 1911 provincial election. The eagerness of the Cape Breton socialists made Roscoe temporarily forget the problems the party was having elsewhere. "Glace Bay local has the best bunch of well-informed socialists that I have ever met," Roscoe wrote after the first of two trips to Cape Breton, "and the campaign that they are carrying on at present is absolutely free from vote-catching tricks of any and all kinds."[1]

The party's candidate was Alex McKinnon, Glace Bay's town engineer and the first Socialist Party candidate east of Saskatchewan to run for election. McKinnon was at home among the miners—he had fought his way out of poverty by working in the mines. Later he had attended night school and a U.S. university to get his engineering degree. Even though there was little chance of winning the election, the socialists believed that running a candidate was good promotion for the movement. They conducted a house-to-house canvass of the riding, and Charlie O'Brien, the Alberta MLA, visited the riding in support of the campaign. Roscoe's only criticism was that McKinnon and J.B. McLachlan gave speeches criticizing the individuals who ran the coal industry instead of pointing out that the system of capitalism was responsible for many of the miners' problems.

On election day McKinnon polled a total of 713 votes, mostly in the mining towns where there were Socialist Party locals. The riding was a two-member constituency in the provincial legislature, and the Conservatives and the Liberals each ran two candidates. McKinnon's vote amounted to 11.4 per cent of the total cast for the leading candidate, Conservative John C. Douglas.

In Saint John, where efforts to run a candidate in a 1911 election were less successful, the Socialist Party local had a permanent meeting place where it held a propaganda meeting every Sunday evening. The most active member of the Saint John local was Fred Hyatt, the articulate, class-conscious British socialist whom Roscoe had briefly worked with in Calgary when they were organizing the poor. Hyatt immigrated to Canada in 1904 and had toiled at an iron works company, as a harvester, and as the Socialist Party organizer in Calgary. He had moved to Saint John in 1910, took a job shovelling coal on the docks, and was soon an influential member of the International Longshoremen's Association.[2] He also quickly became the organizer of the Socialist Party local. Other leading members of the Saint John local were J.W. Eastwood and Alec Taylor, both Brits, and Colin McKay, who later wrote many articles on socialism and politics for a number of Maritime publications.

The socialists got the idea of running a candidate in a by-election in Saint John after they were told they would no longer be allowed to hold meetings on the city's streets, even though the Salvation Army could continue to hold its gatherings. The ban on the socialists was supported by J.D. Hazen, a former New Brunswick premier. When Hazen announced his intention of seeking a seat in the House of Commons in a federal by-election to fill a position in the cabinet of Prime Minister Robert Borden, the other parties stepped aside and said they wouldn't be running candidates against him. This withdrawal infuriated the socialists, who wanted to see Hazen challenged on the free speech issue. The Socialist Party announced it would run Fred Hyatt.

When Hyatt and his official agent arrived at the court house on nomination day, they found the doors locked. According to Hyatt, they banged on the door for twenty-five minutes, only to have people inside occasionally poke their heads out a window and laugh at them. They finally gained entrance by a back door and located the returning officer. On the advice of the city recorder—who was also the head of the Conservative Association—the returning officer rejected their nomination on the grounds that they had made a minor error in filling out the name of the riding. Frustrated and angry, the socialists had no time to redo the papers. Hazen was elected by acclamation.[3]

The decline of the Socialist Party in most areas of the Maritimes was a severe blow to Roscoe. He had given four years of his life to the struggle, and now he found it crumbling around him. Too few people were prepared to join a party that, as Roscoe said, wanted to "turn society on its head." One organizer complained that workers wouldn't even give up the price of a couple of glasses of beer to subscribe to a socialist newspaper.

To a large extent Roscoe blamed the failures of the party on what he felt was the general conservatism of the Maritimes. He wrote about an atmosphere of superstition, bigotry, and medieval backwardness in the region. "There are portions of the country where one can imagine himself set down in Europe during the dark ages," he said bitterly. "Religious superstition is rampant. The old fables and the 'Divine right of kings' are explicitly believed in by a very large majority of the people."[4]

Many Maritimers had good reason not to support revolutionary socialism. In most communities, business pressured people to stay away from the socialists. Some of the Socialist Party's key workers had been blacklisted because of their socialist activities. Roscoe's efforts to continue building a party chapter in Albert failed when the village establishment decided that the Socialist Party should be boycotted. When Roscoe had visited Cape Breton in July 1910, he had found that three Glace Bay party workers were blacklisted. When Gribble was speaking in Stellarton, Nova Scotia, the intimidation of business owners had kept employees from attending the socialist meetings. Intimidation was an especially effective tool against prospective party members who had families to support. The strong bias against socialism was reinforced by both the church and the mainstream press.

Later it occurred to Roscoe that one of the reasons why people in Albert—and possibly throughout the Maritimes—were afraid of his ideas was because he attacked their belief in the sacred nature of property. "These were not wealthy people," Roscoe wrote in his unpublished autobiography, "a few were well-to-do and they interpreted socialism to mean that their property, for which they had toiled a lifetime, would go into a pool from which the lazy ne'er-do-well could live at ease. I was really trying to show them that ne'er-do-wells as the heads of monopolies were living on the fat of the land at the expense of all of us."

While some people agreed with the attacks on capitalism, they felt that the alternative offered by socialism was too vague and untested. Most people were concerned about earning a living and getting ahead in the world—not with fighting for some sort of Utopian society. Some people said they wouldn't join a party that had no platform and no immediate demands. Confronted on this issue, Roscoe wrote, "No, we haven't yards of immediate demands tacked onto our platform and we don't propose to tack any on. We are not catering to cowards...nor do we intend to. Our aim is to make economists of the workingmen."[5] As Roscoe found out, this wasn't good enough for most people.

Roscoe believed that the economic and social structure of the Maritimes worked against the development of socialism. The strongest support for the socialists was in areas where the party was able to organize among the coal miners and factory workers. In other parts of the Maritimes people earned their living from small-scale farming, lumbering, and fishing. Roscoe believed that organizers were at a distinct disadvantage because the individualistic nature of this work made people feel they were dependent on their own resourcefulness to earn a living. Often at the centre of economic control in the Maritime towns and villages were local, conservative businessmen and, because people were closely integrated into these communities, they were less likely to be radical than the workers Roscoe had come across in western Canada who belonged to large unions.

❧

During the time he worked for the Socialist Party, Roscoe's personal life had gone through a major change. While he was in Springhill for the miners' strike he met Margaret Munroe, a niece of socialist Seaman Terris. Margaret's family was a part of the coal mining community; as a youth her brother had worked in the mines. Only fifteen years old, she was still in school and living with her mother, who owned a boarding house where several of the miners lived.

Roscoe, at age twenty-two, became infatuated with Margaret, who was pretty and vivacious. They had what he later described as "one of those sudden romances." Roscoe, who had little experience with women to that time, began courting Margaret during the winter of 1909-10 and revealed the romantic side of his nature by writing several love poems for her.

Margaret couldn't have been more different from Roscoe. Where he was reserved and serious, Margaret was youthful, full of giggles, and a bit flighty. Roscoe had travelled to the United States and western Canada, while Margaret had lived a sheltered life in Springhill, never travelling further afield than Moncton.

Margaret thought that Roscoe was a dashing young man who was quite unlike the miners she encountered at her mother's boarding house. She was impressed by his sincerity and his compassion for the striking miners and their families. She liked the way he would sit for a long time and talk with her. She also liked his expressive eyes and the way he looked when he dressed up in his good suit. Given the seven-year difference in their ages, she undoubtedly looked up to him as a wise and worldly figure.

In August 1910, at the height of the strike, a month before Margaret's sixteenth birthday, she and Roscoe were married in Springhill. No one in the family seemed particularly concerned about her youth. Margaret's mother, Flora, approved of the marriage. Margaret was not doing particularly well in school and her mother may have thought that marriage was preferable to having her daughter live at home and work at the boarding house.

The fact that Margaret didn't seem to be overly curious about the workings of society at large or especially interested in socialism appears not to have bothered Roscoe. Like almost all men in the socialist movement and outside, he had a traditional view of the role of women.[6] Although he admired the intelligent, independent women who were active in the socialist party—women such as Sophie Mushkat and Fanny Levy of Moncton, Lois Brison of Halifax, and Marion Palmer Purves of Douglas Harbour, New Brunswick—he expected his wife to look after the home while he went out into the world to earn a living and seek intellectual stimulation. Margaret too seemed comfortable with such a relationship.

The newly married couple took up residence in Albert, where they lived in a small house the nursery purchased for $1,100. Roscoe's half-sisters, Clara and Mabel, were amazed at how young Margaret looked when they met her for the first time. They thought she was more like a schoolgirl than someone's wife. She was jolly and funny and always laughing. She quickly became popular with Selina and the rest of the family.

❧

Four years in the socialist environment had changed Roscoe. His active role in the party had helped him become a much more confident and capable person. He was proud of his contribution to the socialist cause. He began to wear a bright red tie to events and socialist meetings, and he closed his letters with the slogan "Yours in Revolt." Wilfred Gribble saw noticeable changes in Roscoe's organizing and speaking abilities during the few months the national organizer had spent in the Maritimes. "Comrade Fillmore has developed wonderfully," Gribble wrote. "His limit [before] was about a halting five minutes, but now he can hold forth unhesitatingly at good length. He has a good clear voice that carries splendidly.... More than all, he delivers the right goods."[7]

The time Roscoe had spent organizing allowed him to develop his own ideas on how the socialist movement should be run. He came to adopt a forceful, revolutionary stance. He was an idealist with no time for the pragmatists who predicted that revolutionary socialism would not be successful. He had become one of the most prolific writers in the Maritime socialist press, publishing articles in *The Western Clarion*, *Cotton's Weekly*, and *The International Socialist Review*. Between 1909 and 1911 he wrote more than fifty feature articles. His output rivalled that of H.H. Stuart, who to that time had published more articles on socialism than any other Maritimer.[8] Roscoe also wrote short stories and poetry on topics such as the Springhill miners' strike and Karl Marx. While at least two of his poems were published in the socialist press, most of his creative writing was undistinguished.

Roscoe's writings tended to be the vigorous polemics typical of the socialist press of the era. Expressing a good amount of bitterness and anger, he would argue forcefully that the capitalist class had to be overthrown. In one article in *Cotton's Weekly*, after explaining in detail how capitalism exploited many workers, he wrote:

> Do you wonder that we hate the capitalist class? Do you wonder that I sometimes wish I possessed the strength of Sampson, that I might slaughter these vampires, these bloodsuckers? I would enjoy pulling their marble palaces about their ears as they revel on the proceeds of our misery.[9]

At one point Roscoe came to the conclusion that it might be necessary for working people to defend their interests by armed struggle. Once working people understood how they were being exploited they would want to elect socialists to office, and if capitalists tried to interfere it might be necessary to use rifles to get justice. "We must be prepared to meet the capitalist class with its own weapons," he wrote. "Ballots opposed to ballots. Bullets to bullets if necessary."[10]

Roscoe prepared himself for the possibility of armed conflict by joining the militia—the 74th Regiment, New Brunswick Rangers—and taking part in summer military exercises at nearby Sussex, New Brunswick. There militiamen were housed in tents, wore red tunics and blue trousers, and concluded their training by staging a huge sham battle that was a great attraction for the local citizens.[11] While these militia camps were regarded by many young men as a chance to have a booze-up, Roscoe took the rifle training and drill practice seriously. But when he saw the militia being used against workers by the government and companies to prevent and break up strikes, he soon concluded that armed intervention against the capitalist class would not work. He noted that in schools students were taught militia training hand in hand with the values of capitalism. The radicals were not likely to be able to wrest control of the military away from people who had been so indoctrinated.

Roscoe's strong belief in revolution came in part from his own background. His earliest childhood memories were of the narrow-minded Baptist philosophy that had dominated the life of his family in Lumsden. As he saw how religion controlled the lives and thoughts of many of the Maritimers, his antipathy toward the church increased. He combined his dislike of religion with his hatred of capitalism, writing that priests and preachers were "the accomplices of cut-throat financiers and labour skinners" such as John D. Rockefeller, Andrew Carnegie, and J.P. Morgan.[12]

The antagonism that had developed between Roscoe and Rev. Snelling of Albert led Roscoe to leave the Baptist Church officially in 1911, although he had not been active since he was a teenager.[13] Thereafter he described himself as a "materialist monist," a philosophy that encompassed a holistic view of the world and rejected a dichotomy between the spiritual and the material or, for that matter, the political and the economic. For the materialist monist, according to writer Helena Sheehan, "The belief that the economic, political, cultural, and intellectual spheres were separate and independent of each other was the most fundamental of illusions."[14] The philosophy was influenced by many currents in nineteenth-century thought, including Marxist materialism, Engels's view of the dialectic of nature, and Darwin's theory of evolution. Roscoe believed that history was determined by economic processes, and he agreed with Marx that human history was the history of class struggle. He also seems to have been influenced by Engels's sense of "how totally our origin and destiny were bound up with the rhythms of the natural world."[15] Despite his strong dislike of religion, Roscoe did not speak out against anyone in his family who attended church. When the Socialist Party debated the merits of the church in 1911, Roscoe wrote that religion should be a private matter.

Roscoe's radical stance was also connected to the early death of his mother from tuberculosis. If the family had not been poor, Roscoe believed, they could have moved to the western United States and his mother might have been cured. He blamed the capitalist system for her death, and as a socialist he wanted to do whatever he could to see that other women did not suffer a similar fate.

In one of his articles Roscoe wrote about a day when he had seen a timid and worn woman no more than forty years old standing in front of an Albert store. The sight brought the painful memories of his own mother's suffering back to him. He knew that if his mother had lived she too would have been worn down and sick from a life of poverty just like this woman.

> Born on the farm, perhaps she came to the city when fifteen or sixteen years old full of joyous hopes for the future. She had been taught that woman's mission is in the kitchen.... So she set to work angling a man. In the course of time she caught one. For twenty years or more she had been a workingman's wife or perhaps slave is a better term. She has borne and reared several sons and daughters who are in their turn wage-slaves or slaveys.... For years she has borne the drudgery of the kitchen. She has walked the floor at night with the sick baby and then when morning dawned bent over the wash-tub all day. She has received the kicks and brutal language which her tired husband should have lavished upon his boss. And now at the age of forty, she is the worn-out slavey of a broken-down wage-slave. Isn't it a great system that crushes the mothers of the race and makes beasts of burden of them?

For Roscoe the very fact of the woman's poverty was a justification for socialism, and he was angry that people in general did not rebel over the existence of such exploitation.

> If my mother was alive and had been crushed like the working woman whom I saw that day, I would not remain acquiescent. I would smash the fellows who are responsible in every possible way. I would raise hell before I would see my mother suffer like that. And I said to myself that day, "I am glad my mother died years ago. I am glad she is not alive to be crushed into a beast of burden."[16]

Although his instincts were kindly and he had a strong compassion toward people trapped in poverty, Roscoe could be impatient toward those who, for whatever reason, did not agree with him about the power of socialism to deliver them from their misery. In some ways he was more conservative than he would have liked to admit. Despite what his writing revealed of a sharp understanding for the oppression of women in the home, he maintained a traditional marriage in which he was very clearly head of the household. He was a strong believer in the role of the family and criticized the "apologists of the master class" who argued that socialism would break up the home.

Whenever he could, Roscoe pointed his finger at the greedy people in society who prevented the poor from getting a little security. When proposals were made to establish a universal old age pension, many well-to-do people argued that the promise of a pension later in life would ruin the character of the workers and take away their incentive. Roscoe, who apparently had been reading about salaries paid to civil servants and the judiciary, questioned the logic of the argument by citing the situation of superior court judges in Quebec.

> The worker and the judge are both human beings and both subject to the same reaction to the same environment. When a judge gets $7,000 a year do the people say that the beggar should be frugal and save up for his old age? Not a bit of it. The judge is supposed to be a free spender. He is supposed to "live in a manner befitting his station."... When the judge gives a wine party to his friends the finger of scorn is not pointed at him. On the contrary, the people declare what a fine, generous chap the judge is. When the judge has done his twenty years just outside the jail doors and he is entitled to free board and keep to the extent of $4,000 or $5,000 a year from the nation's pocket, does the judge wax indignant at the thought of being fed at the public expense? Not a bit of it. He is anxious for the day to come when he can eat without work. When he gets his pension his friends congratulate him. He smiles and thanks them for their solicitude and declares that he is glad the long pull is over. Never a word about his being pauperized. Never a word about his not having saved enough to live on.[17]

❧

While the small Maritime socialist movement was struggling, a crisis was brewing elsewhere in the country. The Ontario chapters of the Socialist Party had originally accepted the party's radical doctrine, but when socialist candidates lost at the polls members began to have doubts. The Ontario chapters demanded the inclusion of immediate demands in the platform, a less antagonistic policy toward trade unions, and a national convention to end the autocratic control of the Vancouver leadership. When the Dominion Executive refused to listen to these demands, the party began to disintegrate.[18] Ethnic groups also expressed dissatisfaction with the party. The Dominion Executive, remaining firm, expelled many of the disgruntled members.

By early 1910 a group of Winnipeg socialists who had left the Socialist Party were laying the groundwork for a new, more moderate political party. By December 1911 the new party—the Social Democratic Party—was officially formed, with headquarters in Ontario. The revolt badly damaged the Socialist Party: It lost many members to the new party, its revenues were slashed, and, by early 1912, *The Western Clarion* was unable to maintain its regular publishing schedule.

Signs of the larger crisis in the Socialist Party were evident in the Maritimes. The party began to take action against members who were not loyal to the group's hard-line positions. Immediately following the federal election of 1911, the Maritime executive expelled two Glace Bay members accused of working for non-socialist candidates in the election.

Roscoe became involved in the campaign to defend the party's purity, leading a fight against two of his Cumberland County associates, Jules Lavenne and Adolph Landry. Roscoe argued that the pair had helped the Conservatives during the election. Mayor Tom Lowther of Amherst, a Liberal Party supporter who was opposed to the socialists, told a meeting that Landry had offered to work for the Liberals during the campaign if they would pay him a thousand dollars. Then, according to Lowther, Landry said if the Liberals weren't interested he would work for the Conservatives, for the same price. Lowther pointed out that it seemed that Landry had got the money he was seeking, because he was speaking on behalf of the Conservatives.[19] It appears that Landry, who became a prominent Maritime rights advocate in the 1920s, made no attempt to defend himself in the socialist press.

In a similar charge Roscoe said that Lavenne was guilty during the election campaign of "occupying the same political couch" with Nathan Curry, president of the Canadian Car and Foundry Company of Amherst and of the Manufacturers' Association of Canada.[20] Accusing Roscoe of employing the tactics of the capitalist press, Lavenne wrote that he did not know Curry.[21] At the next meeting of the Maritime executive it was stated that Lavenne had admitted being paid to work for the Tories and that he had received a certain sum for his vote. Despite his denial, Lavenne was expelled by the party in January 1912. Lavenne still considered himself a good socialist and continued to champion the causes of working people.

By 1912 the Social Democratic Party (SDP) had established chapters from Quebec to British Columbia, which left the Socialist Party reeling. The party couldn't generate enough subscribers for *Cotton's Weekly*, which had been committed to the party until that time. Hurting financially, editor W.U. Cotton tried to get the two socialist groups to mend their differences. He failed, and in early 1913 the paper moved to Ontario and began to work more closely with the Social Democratic Party.

For Roscoe and the small group of radical Maritime socialists, the defection of *Cotton's Weekly* was a devastating blow. *Cotton's* was by far the most popular of the radical periodicals in the Maritimes. The main propaganda vehicle in the region now became a much leaner version of *The Western Clarion*. For the first time since he had joined the socialist movement, Roscoe was discouraged. He was also cutting back a little and feeling somewhat guilty for it: In the winter of 1913, when he was in Albert at the nursery helping his father prepare apple trees for spring planting, Roscoe wrote to the Dominion Executive in Vancouver and apologized for not having done a lot of work for the party recently.[22]

Even though he wasn't on the road organizing, Roscoe was firm in his support for the old Socialist Party. He snootily referred to the SDP as "pseudo-Labour-Socialist sentimentalists" and urged those people still true to revolutionary socialism to stick to their principles. "Any attempt to 'swap' the movement for the sake of the votes of reformers, single taxers, Orangemen, Christadelphians or vegetarians, is put down with an iron hand. Result—a small movement it is true, but a movement as solid as reinforced concrete. A steel organization for the workers to rally around when 'the day' dawns."[23]

Roscoe was nearing the end of the first phase of his involvement in radical Canadian politics. While he was frustrated by the movement's failures, he felt he and others had experienced at least some success in their fight to improve the quality of life for working people. They had exposed social injustices and the power of the elite and had pressured for a more co-operative, compassionate society. Socialists made workers aware that there were radical alternatives, and people like Roscoe Fillmore felt that the "captains of industry" would never again rest easy. While the radicals did not achieve the revolution they had hoped for, the exposure of these inequalities made an important contribution to the process of social reform that was under way in Canada.

Meanwhile, Roscoe's personal life was about to change again. While he had been travelling all over the Maritimes preaching socialism, Margaret had given birth to their first child, Richard, in March 1912. Another child was expected in the fall of 1913. Roscoe now had the responsibility of supporting a family. There still wasn't enough work for a full-time job at his father's nursery in Albert—Willard had help from Frank and his brother Arthur. Roscoe knew he would have to look elsewhere for a job.

EIGHT

Apple Farming, War, and a Revolution

The St. John River winds its way through central New Brunswick past farmland not unlike that of Nova Scotia's prosperous Annapolis Valley. In the 1780s Loyalist families had established large farms along the river in dispersed communities such as Burton, Maugerville, and Oromocto; but by the early twentieth century most families had tired of country life and abandoned their farms in favour of life in the cities and towns. Soon after the turn of the century, three New Brunswick businessmen began buying up the abandoned farms along the river with the idea of establishing a huge apple business. By 1911 they had assembled nine hundred acres at Burton, twenty kilometres south of Fredericton, and the company, named the Saint John Valley Fruit and Land Company, had planted about seven thousand trees on this huge farm.[1]

In spring 1913 the fruit company was looking for someone to take over management of the farm. Roscoe Fillmore's experience at his father's tree nursery and the nurseries in Amherst and Rochester qualified him for the job. He was hired as farm superintendent of what would soon become the largest commercial orchard in New Brunswick.

Getting the job was a relief for Roscoe, who now had a family to support. He wasn't afraid of the hard work, but he knew his passion for socialism would have to be tempered. When Roscoe and Margaret arrived in Burton in late spring they found a setting much different from the village life of Albert or Springhill. Large stretches of countryside separated the large farms and old, elegant farm houses. There was no commercial district, only a small general store run by a family of Irish immigrants. Burton's main landmark was its courthouse, and there was a small school and several beautiful old churches built by the Loyalists. Burton's only other business was a lumber mill.[2] The farm was located on the southwest bank of the river. Apple trees in full blossom stretched as far as the eye could see. Jutting out into the river was a wharf where boats docked to load apples each fall and take them down river for export, mostly to Britain.

The company-owned white clapboard house that the Fillmores would live in was located a hundreds yards or so from the river. It was the finest house Roscoe had ever lived in, with its four bedrooms upstairs and a large dining room and living room downstairs—although there was no indoor plumbing. One of the extra delights was a player piano. Outside, rows of tall evergreens surrounded the house, and there were flowering shrubs and perennials.

Roscoe and Margaret had been married three years. By all accounts they were not particularly close emotionally or intellectually, but they managed to live together and raise a family without too many disagreements—Margaret deferring to Roscoe in most cases. To outsiders the family seemed happy, although it was clear that Roscoe was the boss—not surprisingly given Margaret's age and the times. Five days short of her nineteenth birthday, in September 1913, Margaret gave birth to her second child, Ruth Erma. Richard—nicknamed Dick—was now eighteen months old. Young and not entirely prepared for the responsibilities of marriage and children, Margaret was not a particularly neat housekeeper, and she got little or no help from Roscoe in that domain. But Roscoe lavished affection on his two babies. If Dick or Ruth cried in the night, it often was Roscoe who got up to look after them.

Margaret accepted Roscoe's radical political views but showed little interest in learning any more about socialism than what she picked up from listening to her husband. Her life centred around Roscoe and the children, and she had few interests outside the home. Because she had led a closely confined existence in Springhill, her favourite diversion was family excursions in a car provided by the farm. The children enjoyed the usual activities provided by the New Brunswick countryside. In summer, they swam in the river. In winter, huge areas of the frozen river were cleared of snow for skating and hockey. Some evenings they gathered wood and had a huge bonfire.

Roscoe quickly took charge of running the farm. In August he recruited workers from the surrounding countryside to pick apples and pack them in boxes for shipment down the river. The farm also grew small quantities of other fruit—plums, pears, and strawberries. Trees started as seedlings were grown to maturity for planting on the farm or for sale to other farms.

Building on the knowledge of tree grafting learned from his father, Roscoe experimented to develop hardy new apple trees—work that A.J. Turney, the government horticulturalist, would praise as being unique and valuable a few years later.

While Roscoe enjoyed his work, he was aware that he was living a contradiction. He was opposed to the idea of corporations such as the Saint John Valley Fruit Company becoming involved in agriculture. He believed in the preservation of the family farm and knew that many small farms in New Brunswick were seriously in debt. In an article written around that time he expressed the fear that in the same way corporations built huge monopolies in other industries, they would invest in new technology that the small farmer could not afford, putting family farms out of business.[3] His boss Arthur Slipp, one of the company owners, was a lawyer and Conservative member of the provincial legislature—standing for just about everything Roscoe had fought against as a socialist. Slipp was a part of the powerful Conservative Party clique that dominated politics in New Brunswick. A wealthy member of a high-priced city law firm, he had been a Fredericton alderman for several years before being elected to the legislature.[4]

When Roscoe arrived in Burton, Slipp and his law partner, R.B. Hanson, were among a small group of Tories in Fredericton desperately trying to save the career of Premier Kidd Flemming. The premier had been charged with extorting money from companies—the money had found its way into the coffers of the Conservative Party. The Tories failed to clear Kidd and he was forced to resign after a government commission of inquiry.[5] The Conservative government, which later included Slipp as Minister of Lands and Mines, continued to be riddled with problems of embezzlement, payroll-padding, accepting money from liquor dealers, and patronage. In the early stages of World War I, only days after Britain had declared war on Germany, the New Brunswick government proudly announced that it would contribute $75,000 worth of potatoes to the British war effort. It was soon discovered that the administration was buying all the potatoes from loyal Tory farmers, who were making a handsome profit from the governmental generosity.[6]

Slipp and his family had their own cottage down by the river, where they spent much of the summer, so they frequently visited the farm. Slipp would arrive properly dressed in a shirt and tie under a sport coat and wearing stylish "breeches" that went down to just below his knees. A large, sometimes boisterous man, Slipp was popular with the farm hands, but Roscoe, understandably, soon took a dislike to him. He felt his boss wasn't really a friend of the working man.

Despite his bitterness, Roscoe kept his political feelings to himself and concentrated on his new job. He had little time now to devote to organizing for the Socialist Party, but he had not given up his commitment to socialism. He was able to send small donations of $1.25 or $2 every couple of months to the Socialist Party to help keep *The Western Clarion* alive. Roscoe was excited and inspired when he read *The Western Clarion*'s published excerpts from Jack London's novel *The Iron Heel*, a futuristic portrayal of a society where fascism had triumphed. He occasionally made trips to Saint John, where he visited friends who were active in the party.

Roscoe learned that one of the people he had encouraged to become active in socialism, Sophie Mushkat of Moncton, was working in western Canada as an organizer for the Socialist Party. In *The Clarion* Roscoe read that during a five-week period Mushkat spoke to more than forty meetings in the small towns and mining camps of Alberta and British Columbia.[7] Roscoe's younger brother Frank had also become active in the party. On a western harvest excursion in 1913, Frank had started subscribing to *The Western Clarion* and writing letters to the socialist press. Back in Albert the next summer, Frank helped to arrange a second visit to Albert County by Alberta socialist Charlie O'Brien, who was on a cross-country tour. Frank reported in *The Clarion* that, perhaps for the first time, a prominent socialist spoke in Albert without being harassed and insulted.

At the same time that Roscoe and his family were establishing themselves in Burton, international tensions were building to the point where war seemed inevitable. Roscoe, strongly opposed to the brewing conflict, wrote that if war came it would be because of squabbling among capitalists for control of international markets. A month after war had been declared, *The International Socialist Review* published an article by Roscoe in which he argued that European countries had been trying for years to isolate the Germans because they were beating the capitalist countries in capturing world markets for a number of important commodities.[8]

❧

When Britain declared war on Germany at the beginning of August 1914, Canada came to Britain's aid and started to contribute heavily to the war—much to Roscoe's disapproval. More than six hundred thousand men were eventually enlisted in the Canadian army corps. In New Brunswick the war was greeted with excitement and anticipation. The *Saint John Globe* said that young men were "quietly though joyfully" accepting whatever part they could play in the war effort.[9]

For Roscoe and others like him, this was an extremely demoralizing time politically. He wrote in *The Western Clarion* that the war was probably the greatest crime ever committed by capitalist countries against the workers. "Today practically all the 'civilized' nations of christendom are locked in a death grip," he stated. "Millions of men, all (or practically all) of the slave class, are in the field engaged in the pleasant pastime of shooting holes through each other. All to the glory of that glory beast: capital.... We of this supposedly enlightened age are to be treated to the most appalling orgy of blood known in history."[10]

His bitterness was still evident a year later, in a 1915 article in *The International Socialist Review*. "What the hell do we care about the killings in war?" Roscoe asked. "Aren't we killed in droves and armies every year in industry, and just as unnecessarily as in war? Aren't the police forces of New York and many other cities on this continent being drilled in the use of machine guns to mow down those who have escaped the railway smashes and mine and powder mill explosions and feel the spirit of revolt stirring? Killings are as common as cabbages in peace as well as in war. We workers who are in revolt object to war because it makes us pawns, puppets, cannon-fodder at the pleasure of the class we hate."[11]

War undid many of the gains made by working people, Roscoe wrote. "It makes cave-men of the best of our class. It arouses the blood lust that is our heritage from savage, animal ancestors. And it will take us several generations to wipe out by education the racial and national hatreds between the workers of the countries involved that have been aroused to satisfy the masters' lust for profits. It is injuring our class and our chances of successful revolt for a long time to come."[12]

Indeed, the outbreak of war did turn out to be a devastating blow to international socialism. In Western Europe, prominent socialists put patriotism ahead of their political beliefs. Roscoe felt betrayed by the German Social Democratic Party, in particular, which did little in its country to try to stop war from being declared. Roscoe had expected the Social Democrats to establish the world's first socialist government in Germany, and he believed that they should have resorted to armed revolt to oppose the war.

The socialist movements in Germany, France, and Belgium were all taken over by what Roscoe labelled "so-called respectable bourgeois radicals and reformers." Two prominent German socialists who opposed the softening of the socialist movement and the war, Rosa Luxemburg and Karl Liebknecht, became heroes to Roscoe for the leading role they played in the establishment of the radical socialist Spartacus Party during the war. Luxemburg was a brilliant writer and orator, and when Margaret gave birth to a third child, a girl, on December 2, 1915, the baby was named Rosa Luxemburg Fillmore.

During the great upheaval in Germany during the war, the Spartacus leaders, including Luxemburg and Liebknecht, were hunted down and murdered. Roscoe hung huge pictures of Luxemburg and Liebknecht in the living room of the house in Burton.[13]

The answer to the serious setback of the world socialist movement, Roscoe believed, was to support more radical socialist positions. His response to what appeared to be the collapse of international socialism was to write two articles in *The International Socialist Review*, "How to Build Up the Socialist Movement" and "Keep the Issue Clear." He wrote that there was too much confusion over what was true socialism. "Australia and New Zealand were 'won' for socialism of the same brand some years ago and today they are as loud in their jingoism and their 'love of the H'Empire' as any other people. They are forcing conscripts to go to Europe to fight for the capitalists of Britain."[14]

Roscoe decided that committed socialists must take a strong stand. "I believe it necessary that we who are still unmoved by the exigencies of capitalist commerce and consequent world war should get back to first principles and endeavour to found our next international upon the solid rock of the class war and a thorough understanding of society as at present constituted. In order that we may understand, we must think as proletarians, not as pro-German or pro-British bourgeoisie."[15]

In Canada the Socialist Party and its rival, the Social Democratic Party, both opposed the war, and both faced a rapid erosion of membership and support as a result. There was barely enough money to keep *The Western Clarion* running on a monthly basis. The Trades and Labor Congress of Canada, which had adopted a strong antiwar position in peacetime, soon changed its position to support the British war effort.

The outbreak of war led to divisions within the Socialist Party's local in Saint John. With about thirty paid members, the Saint John local was the largest in the Maritimes, with the possible exception of Glace Bay. A new local at Whitehead, fifteen kilometres north of Saint John, had ten members. When Fred Hyatt, the key organizer of the Saint John local, became involved in the campaign to enlist men to go off to war, his loyalty to Britain took precedence over his loyalty to revolutionary socialism. One angry Saint John party member said that Hyatt "has donned a shambles suit, renounced his principles, and played false to the interests of the working class."[16]

❧

Saint John was the place in New Brunswick most wrapped up in the patriotism of the war. The city was known as the Loyalist City for the thousands of people who had settled there after leaving the United States following the War of Independence, and there was great support for Britain. In October 1915 Saint John staged a huge recruiting rally. Union Jacks and war posters were displayed everywhere, thousands of people assembled on the streets, and bands playing patriotic songs led parades to an arena for a giant rally.[17]

But at least one British-born socialist who was living there in 1915 was not in favour of the war or of young men being pushed into conscription. Wilfred Gribble, the socialist organizer who had toured the Maritimes in 1909, had settled in Saint John to practise his trade as a carpenter. In December Gribble attended a socialist meeting in Saint John and made several critical comments about the war that offended one George Worden. Worden went to police and Gribble was charged with making seditious statements.

Gribble's trial was held in January 1916 before Judge James Crocket and generated considerable interest. Crocket was influential in the Tory party and a proud Loyalist. The case—believed to be the first of its kind in New Brunswick—must have been important to the government, because it was prosecuted by the attorney-general, John Baxter. In court the complainant Worden was the only witness called to testify against Gribble. He said that while speaking to the socialist meeting Gribble had made fun of the slogan used to recruit young men for the war: "The King and Country need you!" Worden testified that Gribble changed it to say, "The King and Country will bleed you!" According to Worden, Gribble also yelled, "King George is a puppet!" and said a "person might as well be a German slave as a slave under British rule."[18]

After hearing the testimony Judge Crocket addressed the jury for more than an hour, leaving no doubt in anyone's mind about his views on the case. "It is greatly to be regretted that in these critical and momentous times, when the thoughts of all loyal subjects of His Majesty have been turned to King and country, and when the best men and women, too, are rallying all the young men to fight the greatest peril we have ever faced, there should be uttered in the city of St. John words that are the occasion of this indictment —the first that I know of ever preferred in this part of the country. As it is the first, I hope it will be last in the history of this province, whose founders have left us such an inspiring loyalty."[19] The jury retired and returned an hour later, with a verdict of guilty.

When court resumed a week later for sentencing, Gribble's defence lawyer requested that the case be thrown out on the grounds that the judge had made four serious errors during the trial, including an error of fact when addressing the jury. Judge Crocket refused the request. In a lengthy statement before handing down the sentence he argued that the offence was serious because the comments were spoken in Saint John, a community "in which unflinching attachment and devotion to the British Crown stands out as a predominant characteristic of the whole body of the population." But, said the judge, he would be lenient, because it was obvious that Gribble's judgement had been affected by his belief in socialism—"those pernicious, false expositions of doctrine or belief, with which most unfortunately you have been feeding your intellect."[20] He sentenced Gribble to two months in jail. Gribble began serving his sentence, but the Trades and Labor Congress appealed to the federal justice department on the grounds of Gribble's poor health, and he was released early.[21]

❧

Like many of his socialist friends, Roscoe believed that a socialist revolution would someday sweep the world, but he had no idea when it would come. He thought that some day a desperate worker would throw a brick at a gunman who was trying to crack his head, and the revolution would be on. "We don't know anything about the coming revolution," he wrote in *The International Socialist Review* in the fall of 1914. "It may come all unexpectedly at the close of the war. If, as seems likely, the golden age of capitalism is yet to come as a result of the war, then the social revolution is perhaps a century in the future—perhaps several of them."[22]

No one was more elated than Roscoe when, less than three years later and with the war still raging, revolution broke out in Russia. And the momentous event began with a protest that at first seemed insignificant. Some eighty-thousand metal and textile workers in Petrograd went on strike on February 23, 1917, and the protest—partly occasioned by severe shortages of bread and coal—spread to other workers who took to the streets. When key elements of the military abandoned the Czar in favour of the workers, Russia was in the midst of a full-fledged workers' revolution. On March 15, the Czar abdicated.

The events in Russia amazed Roscoe. But once he had time to think about what was taking place, he was not particularly surprised. Ever since he had been in school he had been aware of the terrible suffering of the Russian people, and he believed that if the conditions existed anywhere for revolution, they existed in Russia. There had been a serious depression at the turn of the century. The country had faced a humiliating defeat in a war with Japan in 1905, in which hundreds of thousands of soldiers were killed. A working-class revolution in 1905 had been defeated. The Great War saw ill-equipped, poorly fed Russian soldiers sent off to be killed. By 1916 public confidence in the government had evaporated.

Roscoe, dependent on the Fredericton *Daily Gleaner* and other newspapers for all his information about the events in Petrograd, became obsessed with getting news from Russia, obsessed about the question of whether a truly revolutionary movement would take hold in Russia or whether a government would emerge that would be just another front for capitalist interests. "I've no doubt I became something of a damned nuisance to my friends and neighbours, most of whom resented being presented with issues that might stir them out of their self-complacent lethargy," he wrote later. "They preferred not to know. Of course, I was an extremist, a fanatic—I freely plead guilty to this." He started to resent Sundays, because no paper came out that day. At first the establishment press, more concerned about seeing Russia continue the war than it was about the welfare of the Russian people, welcomed the revolution. A front-page headline in *The Daily Gleaner* announced, "Russia's Emancipation is Hailed as Epochal Event."[23] But as the upheaval in Petrograd continued, and when it became apparent that Russia was withdrawing from the war, Roscoe was dismayed by the delay of other countries in recognizing the significance of what was taking place. "The fact that a bloodstained and thoroughly reactionary, medieval regime had gone into the ash-can of history," he wrote later, "was not greeted as it should have been with loud huzzas and rejoicing, except among socialists and what one might call left-wing liberals."

Petrograd was chaotic during the summer of 1917. Provisional governments were formed and fell. Factions struggled to gain control of the country. The most important events occurred in the fall when popular support shifted to the Bolsheviks, the one party that believed the country should be run by the Workers' Soviets that had been established in all major areas. The Bolsheviks won a majority in the Soviets of Petrograd and Moscow and won over the troops in the capital. On 25th October/7th November, Vladimir Ilich Lenin, the Bolshevik leader, proclaimed the slogan "All Power to the Soviets," seized power, and formed a new government that would be a dictatorship of the proletariat.[24] Roscoe was jubilant. He wrote later, "I was convinced that events were taking place that would change the world, indeed were changing the world irrevocably at that very moment."

The western Allies did all they could to sabotage the revolutionary

Bolshevik party in Russia. In one of the most unlikely occurrences of the war, British and Canadian military officials conspired to keep Leon Trotsky, the revolution's chief organizer, a prisoner in a Canadian jail at Amherst early in the 1917 revolution. Trotsky and his family were on a ship travelling from New York to Petrograd in March 1917 when British intelligence officials arranged for their detention during a stopover in Halifax. An outraged Trotsky was taken to a prisoner-of-war camp in Amherst where he was held along with about eight hundred German prisoners. Trotsky's illegal arrest caused a stir among Canadian, British, and Russian officials, and Canadian labour leaders protested his detention. The Petrograd socialists finally pressured the British into releasing him, and he marched out of the camp to the enthusiastic cheers of the German sailors. A make-shift orchestra of sailors accompanied him to the gates, playing "The Internationale."[25]

Inside Russia, all but two or three of the reporters covering the revolution in Petrograd had no understanding of what the events meant. When the wealthy newspaper owners in the United States and Europe finally discovered that the upheaval was a class revolution, they tried to dismiss and discredit the importance of what was happening. In the two years from November 1917 to November 1919, *The New York Times* reported no fewer than ninety-one times that the Bolsheviks were about to fall or, indeed, had already fallen.[26]

After the signing of the peace with Germany in November 1918, three hundred thousand Allied soldiers who were in Russia to fight the Germans stayed to fight the Red Army. The Allied forces—including Canadians—threatened to strangle the Russian Revolution in its infancy. Canadian soldiers were among those who mutinied against their officers because they didn't want to fight the Russians. Nevertheless, the actions of the Allies, coupled with their economic blockade, increased Russian casualties from famine, disease, and civil war to close to fourteen million.[27]

The rise of a revolutionary government in Russia had a profound influence on Roscoe and the Canadian socialist movement. Mass meetings of workers and farmers were held across the country, and the thousands of immigrants who had fled Russia for Canada after the failure of the 1905 revolution were especially enthusiastic. Money was donated to the new Russian government. People who had been in favour of the war changed their opinion. Many Canadian unions began to demand higher wages. The Socialist Party of Canada took on a new life.

Roscoe was again inflamed with the zeal of the revolution. Disturbed by the distortions of the mainstream press, he began to give talks on what he believed to be the facts of the revolution. In February 1919 Roscoe was in

Amherst to give two speeches on "The Truth About Russia." The railway-carmakers union hall was packed for the Sunday meetings, and a lot of copies of *The Red Flag*, the Socialist Party's paper, were sold. Roscoe spoke for more than an hour each time, explaining how he believed that the form of government being established in Russia was the purest form of democracy yet developed. He vehemently denied the stories of murder, robbery, rape, and disorder in Russia that the capitalist press had reported. He said the mainstream press, including the Amherst newspaper, was telling lies about Russia.

In its next three issues the *Amherst Daily News* struck back on its editorial page, stating that Roscoe's source of information was limited, that he was not well read, that he did not know what was going on in his own country. "When he asserts that our Canadian newspapers are not anxious to get all the reliable news they possibly can in regard to the actual conditions in Russia, he simply reveals his own ignorance of the newspaper fraternity.... Canadian and American newspapers would pay thousands of dollars to secure a reliable, unprejudiced report on how the Soviet system is actually working out in regard to industry." The editorials also said that the Russians were funnelling ten-million rubles into foreign countries to promote revolutionary socialism. "It might be pertinent to ask how much of this five million dollars was being circulated in Canada, and what small portion of it was coming Mr. Fillmore's way."[28]

After the third editorial Roscoe responded with a seven-page rebuttal, but instead of publishing the letter the *Daily News* ran another editorial criticizing Roscoe yet again. The paper said it wouldn't publish his letter because the columns of space given to the investigation being held before a Senate Committee in Washington already gave the same publicity to supporters of the Soviet system that it gave to its critics.

Roscoe devoted as much time and energy as he could to promoting the cause of the new revolution, though he was limited by his job and family obligations. Socialists proposed sending him on a two-month tour of the Maritimes to speak on Russia, but he couldn't spare the time. Still he revelled in the success of the Bolsheviks and dreamed of one day visiting Russia to see the revolution for himself.

Roscoe wrote later that he had little doubt that a great deal of social legislation in most countries was due to the influence of the Russian Revolution. "They put new spirit in working people everywhere and in many cases enlisted middle class and well-to-do people in the struggle for more labour organizations and for social reform," he said. He was sure the ruling class and big business everywhere were trembling in their boots in fear that their workers might do as the Russians had done—take over government and industry.

NINE

Red Scare: The Socialist Party of Canada

DESPITE HIS preoccupation with the events of the Russian Revolution, Roscoe Fillmore was doing well as manager of the apple business at Burton. Under his direction the company flourished, shipping hundreds of barrels of apples each fall to British and Canadian markets. Apples and other fruit from Roscoe's orchards frequently won top prizes in competitions in Saint John and Fredericton. Government delegations occasionally toured the St. John Valley company orchards, commenting favourably on the high quality of the produce.

Roscoe's good work won him recognition among other farmers, and he played an active role in the provincial fruit growing industry. In 1917 he was elected president of the New Brunswick Fruit Growers' Association, a government-assisted organization that promoted the apple industry and helped farmers with such matters as pest control and new species development. Roscoe worked within the association to encourage farmers to end their years of isolation and join together to meet growing competition for their exports. He gave talks for other farmers on growing apples and nursery stock.

Back home in Albert, Roscoe's brother Frank was gravely ill with Spanish influenza. Frank, who was twenty-five, had spent the summer of 1918 working in Amherst at W.A. Fillmore's nursery and was back in Albert just a few weeks when he became sick. Thousands of Canadians were dying in a Spanish flu epidemic that was spreading across the country. More than three thousand people died just in Ontario in October 1918. In late October Frank contracted double pneumonia and his condition seriously deteriorated. A few days later he died.

Roscoe was devastated by Frank's death. He had high hopes for his young brother, who was becoming more and more active as a socialist. Roscoe described Frank, who had become interested in socialism at the age of fifteen, as "a thorough going clear cut Red." He went to Albert for the funeral and was pleased to find out that Frank didn't turn to religion on his death bed. "He went as game as one could wish—no crawling to 'Gawd,' no prayers, no fear of 'Eternity,'" Roscoe wrote to one of the western Socialist Party leaders. "Our people are all religious as hell and worry because he was not saved."[1]

There was more bad news. Willard's business was suffering because hardly anyone was planting new fruit trees during wartime. Sales at the Albert nursery shrank to the point where Willard, now fifty-seven, barely had enough income to keep things going. Frank's death meant that Willard and his epileptic brother Arthur were left running the nursery. Despite the difficulties, Willard and Selina managed to send their two daughters to college—something that earlier had not been possible for Roscoe.

A year after Frank died Roscoe lost a second close friend. Clarence Hoar, Roscoe's cousin and the first person to join him in trying to build socialism in Albert County, died of tuberculosis in 1919 in Colorado at age twenty-six. Since being forced out of his job as a bank teller in Albert because of his socialist activities, Clarence had lived in Portland, where he was active with the Socialist Party of Maine. He was a founder of the socialist paper *The Issue*, and his letters and articles had frequently appeared in *Cotton's Weekly* and *The Western Clarion*. Clarence was one of the few Maritime socialists Roscoe had known who had refused to give in to the pressures of business when his employer told him to choose between his job and socialism.

❧

By the end of 1917, radicalism was gaining new strength in Canada. Workers were angry over the sacrifices they had been forced to make to pay for the war. From 1915 on, most Canadians had experienced a drastic erosion in their living standards. Prices increased from between 10 to 20 per cent a year.[2] Meanwhile, workers were aware that most corporations were making record-setting profits. The government gave out contracts for millions of dollars' worth of military supplies, guaranteeing big profits for these corporations. The book *History of Canadian Wealth*, written by Gustavus Myers in 1914 and read avidly by radicals, estimated that "less than fifty men control

$4-billion, or more than one-third of Canada's material wealth."[3] Workers gained inspiration from the fact that the first revolutionary government in the world had been established in Russia, and soldiers were coming home from the war with expectations of a better future in Canada. As the Canadian economy revived and thousands of people entered the work force, there was a major increase in militancy.

In this volatile environment the Socialist Party of Canada had re-emerged from the ashes of its disastrous prewar era to launch a new campaign of revolutionary socialism. A new group of men controlled the party, and unlike earlier leaders they believed it was necessary to work with radical elements within the national trade union movement to build a revolutionary socialist movement. The change in emphasis was extremely important because it would allow the party to capitalize on the increasing militancy of union membership. What followed was the greatest period of industrial unrest in Canadian history.[4]

Even though Roscoe knew there was little chance of starting a socialist revolution in New Brunswick, he again began to work closely with the Socialist Party of Canada. He wrote to the Vancouver office to buy copies of *The Western Clarion* and other literature to distribute. A typical order, sent off in October 1918, was for a hundred copies of the leaflet "Bolsheviki Declaration of Rights," twenty copies of "The Russian Situation" by Arthur Ransome and John Reed, and four copies of *Soviets*.[5] After his day's work on the farm was completed, Roscoe often worked late into the night in his library, writing and thinking of ways he could help the socialist cause in the Maritimes.

Roscoe worked to strengthen the small network of socialists in the Burton-Fredericton-Saint John area. Among them was Fred Thompson of Saint John. Thompson had joined the Socialist Party at sixteen and worked as a labourer on Roscoe's farm. Thompson soon moved on to Halifax, where he wrote for *The Citizen*, a socialist paper edited by Joe Wallace, and became involved in the early stages of the 1920 strike at the Halifax Shipyards.[6] Later that year Thompson moved to Winnipeg and then to the United States where he began a career as an organizer and later newspaper editor for the Industrial Workers of the World. Years later Thompson said that Roscoe was the first socialist he had met who knew his subject thoroughly.

In the Burton area Roscoe interested a number of people in the cause of socialism, including Clarence Martin, Roscoe's foreman on the farm.[7] In Saint John, Sanford White and M. Goudie organized regular Socialist Party meetings, distributed party books and literature, sold subscriptions to *The Western Clarion*, and donated money to the party. Goudie, a music teacher, often entertained friends in his home by playing the violin, after which he'd ask for a modest donation for the Socialist Party. When socialists believed that the authorities were opening their mail, Goudie suggested that comrades place a short hair in any letter they wrote to another socialist. If the hair was missing, they could assume someone had tampered with the letter.

Roscoe set up what the Fredericton *Daily Gleaner* later called a socialist school at Oromocto, a village near Burton, for the people he had interested in radical politics.[8] In truth the school was little more than Roscoe getting together with a handful of socialist supporters and curious onlookers and talking about the history of socialism and why society should change. Despite the fears of *The Gleaner*, most of the farmers who lived around Burton were not interested in radical socialism. Many of them participated instead in the United Farmer Party, a liberal-minded party that was challenging the province's political establishment. One of the organizers of the party was Roscoe's former co-worker in the Socialist Party, H.H. Stuart. The populist Farmer Party won a federal by-election in 1918 and ten seats in the 1920 provincial election, but this gave little encouragement to Roscoe, because he felt sure the farmers were only interested in reforming the existing capitalist system.

❧

The radicalism growing elsewhere in Canada led many people to become convinced that hoards of slogan-spouting leftists were planning sabotage, murder, and revolution to seize control of the country. As the war drew to a close, the country was swept by a wave of anti-Bolshevik rumours that led to a Red Scare of greater dimensions than what Roscoe had experienced during his work in the Socialist Party from 1908 to 1912.

In the fall of 1918 the Canadian government issued wide-sweeping Orders-in-Council banning most of the country's socialist publications and making several socialist organizations illegal. Somehow in the confusion the Socialist Party was not banned, though *The Western Clarion* was. The government move put Roscoe at risk. He had been reading and distributing *The Western Clarion* for eleven years and now it was banned, as were all the publications of Charles Kerr and Co., including *The International Socialist Review*, which Roscoe wrote for and read religiously to keep up on developments in Russia and Europe. Also banned was a long list of books and pamphlets—including *Looking Backward* by Edward Bellamy and *The Communist Manifesto* by Marx and Engels—both of which were in Roscoe's library in Burton.

Roscoe wrote Chris Stephenson, Socialist Party secretary at party headquarters in Vancouver, declaring that he had no intention of destroying his personal library. He sent the party ten dollars, with an offer of more later, to fight the government actions. "Of course they can't stop us," Roscoe wrote. "[Prime Minister Robert] Borden and Co. are damned small crabs to stand in the way of the slaves once they get wise, but the stoppage of *The Clarion* will make our propaganda a trifle harder and slower perhaps. Glad you fellows have been able to keep out of jail so far."[9]

A month after *The Western Clarion* was banned the Socialist Party began publishing a new newspaper, *The Red Flag*. Unable to use the mails to ship *The Red Flag*, the party resorted to secretly sending bundles across the country by express. *The Red Flag* concentrated on publicizing the successes of the Russian Revolution and became the first publication to introduce Lenin's writings to the Canadian public.

The fact that Roscoe was unable to play a prominent part in the socialist upheaval occurring elsewhere in the country led him to resent his obligations. He began to lose interest in the farm and sometimes argued with Margaret over matters that, to her, seemed trivial. In a letter to Socialist Party headquarters in early 1919, Roscoe explained that the war had demoralized him and that he had forgotten about socialism for a while. But, Roscoe wrote, now he had his second wind and the only reason he wasn't out working full-time promoting socialism was because he had a wife and three children. "If I were single, all hell wouldn't keep me on the farm another twenty-four hours."[10]

The mainstream newspapers throughout the country were full of hate material, claiming that the followers of Lenin were a destructive, bloodthirsty horde. An editorial in Fredericton's *Daily Gleaner* said that a collection of reports on Russia put together by the British foreign office showed "a record of bloody crimes, wholesale massacres and diabolical tortures."[11] The collection, said the newspaper, gave hundreds of specific examples of terrible acts committed by the Bolsheviks. A nineteen-year-old girl accused of espionage was tortured by being slowly pierced thirteen times in the same wound by a bayonet, and the bodies of soldiers who had fought against the Russians were found with their eyes torn out and gramophone needles thrust under their finger nails.

The antisocialist propaganda led to attacks on socialists throughout North America. On the first day of a roundup of suspected Bolsheviks in the United States, there were 591 arrests. Canadian officials went to Washington to determine ways that the two countries could work together to root out the Bolsheviks. When Vancouver workers went on strike to protest the killing of Albert "Ginger" Goodwin, a union leader who had been shot while evading the draft, soldiers rioted and broke into the labour temple in Vancouver.

After unsuccessfully trying to throw Socialist Party activist Vic Midgley out of a second-storey window, the soldiers forced him to kneel and kiss the Union Jack.

In Fredericton in July 1919, when a circus was visiting town, Roscoe and another socialist passed out copies of *The Red Flag* to the general public after an evening performance. The newspaper, considered illegal by police because it was a publication of the Socialist Party, carried stories protesting the peace agreement that had ended the war. The next day *The Daily Mail* described how the papers had been given out by two men who stood in deep shadows. Police had tried to apprehend the people responsible but were unable to catch them. The day after that *The Daily Gleaner* reported that there were two nests of radicals with strong Bolshevik tendencies, one in Fredericton and one in the Burton area. The paper said that few, if any, of the local immigrant population had anything to do with the socialists. The Fredericton Labor Council, denouncing the distribution of the Bolshevik literature, said it had received a shipment of similar literature but burned it before it could reach the general public. President Harry Ryan promised that if the council found out who was distributing the literature, the person would be turned over to police. *The Daily Gleaner* reported that the attorney-general's department had been watching the Bolsheviks for several months and had considered laying charges against one of the leaders, presumably Roscoe.

In another incident, Roscoe wrote the proprietor of a Fredericton movie theatre, complaining that a recent picture had depicted Bolshevism in a very negative, inaccurate light. The film was one of several shown around North America that attempted to incite hatred against socialists. Roscoe's letter was turned over to the police. *The Daily Gleaner* observed, "The writer warmly defended the Bolshevik and extolled the principles for which the Reds stand."[12] Promotional material sent to theatre managers for another film, *Bolshevism On Trial*, suggested that theatre managers ask groups to stage antisocialist protests to drum up interest in the film. It also was suggested that red flags be put up around town and soldiers hired to tear them down, and that cartoons depicting Bolsheviks running around with flaming torches, burning factories and raping women, be posted throughout the town.[13]

When he could spare the time, Roscoe promoted the cause of socialism further afield—in Saint John and Amherst—where he hoped people would be more receptive. Through the winter of 1918-19 he went to Saint John about once a month to meet with local socialists and speak at their meetings. He was bitterly disappointed by the results and wrote to party secretary Chris Stephenson: "Saint John is a god-damned discouraging hole. Bolshevism has got their goat and about forty or fifty is the average attendance, most of them members of the party. Practically the whole of organized labour in Saint John attends Mass every Sunday!"[14]

❧

Roscoe frequently took the train to Amherst, where he hoped to rebuild a chapter of the Socialist Party that was not very active. At the same time that the highly publicized general strike was on in Winnipeg, a similar but unrelated crisis was building in Amherst.

During the years since Roscoe had been active as a Socialist Party organizer in Amherst, tensions had been seething beneath the surface. Workers were bitter because the town had lost several industries and hundreds of jobs to central Canada. The town was still home to a variety of industries—such as railway cars, textiles, shoes, boilers, and steam engines—but there had been little reinvestment in product lines and the work force had been disintegrating, declining from about 4,000 employees in 1907 to 2,500 in 1919. Now there were rumours that the Canadian Car and Foundry Company, the town's biggest employer, was going to relocate part of its operation. For a more militant labour force at the end of the war, issues such as improved working conditions and wage parity with central Canada were also becoming increasingly important. In 1918 tensions were heightened by work stoppages at two major companies and the formation of a Federation of Labor, which the workers planned to use to challenge the town's big companies.

After he spoke about Russia in Amherst in February 1919, Roscoe wrote to Stephenson in Vancouver, telling him that there were about forty young, energetic Reds in Amherst and that the town had the best potential for socialism of any part of the Maritimes.[15] Many of the men there had been active in the Socialist Party before the First World War.[16] With their support the Socialist Party was able to re-establish its chapter in Amherst.

With the coming of spring, Roscoe's work commitments at the farm would prevent him from spending much time in Amherst. Frustrated, he may have been looking for a way to leave his job at Burton and follow the political trail, because he wrote Socialist Party officials to say that the Maritimes was ready for a full-time socialist organizer. He complained that there was no one in the Maritimes who would take on the task and argued that with the socialist support existing in Amherst and Cape Breton the post could soon be self-supporting. After asking whether the party had a single man they could send down to the Maritimes, Roscoe said that if the Maritime socialist movement could support him, he would "hit the trail tomorrow."[17] To do this, his family would have to be guaranteed $1,200 to $1,500 a year. Socialist Party organizers in Vancouver didn't take Roscoe up on his offer. Because of the Red scare out west, the party was under siege. Meetings had been stopped across the country. The owners of halls were intimidated by threats of violence. Money was in short supply.

In Amherst in April 1919 a new militancy was evident when the Amherst Federation of Labor decided to affiliate with the new One Big Union, which was being created in the west by leaders of the Socialist Party and other radicals. Partly because of the influence of the socialists, the Amherst labour movement had become extremely class conscious. Instead of organizing by trades, like the conservative Trades and Labor Congress did, the Amherst Federation of Labor included all of the workers in a factory in one local, in much the same way as the IWW had done or the way the One Big Union was now attempting to do. Still, the OBU was then only in the formative stages and not ready to issue charters. The Amherst group showed that it supported the same radical ideas that were sweeping western Canada by distributing OBU literature to its members and adopting the title "One Big Union" for its own movement.

By May 1919 Amherst was set for the largest general strike in the history of Maritime labour. The workers suffered from a painful combination of low wages and high living costs. Workers at Canadian Car and Foundry were demanding pay equal to the level that their four thousand Montreal co-workers had just won in a three-day strike. Management told the men they would have to accept lower wages if the company were to remain competitive. The workers refused to accept this dictum. When the car-works company made an announcement that seemed to circumvent the union, the union walked off the job, marched through town, and held a massive union meeting. On the following day workers at other companies started leaving their jobs, and the general strike was on. Workers demanded a nine-hour day, wage parity with central Canada, and recognition of their Federation of Labor. The federation announced that none of its member locals would return to work until the grievances of all locals were resolved.

At first the companies refused to deal with the new Federation of Labor, but as the days passed they began to come to terms. After three weeks all but two of the Amherst firms had settled. Through their solidarity, the workers of Amherst won concessions previously thought to be impossible. For instance, union employees at Robb Engineering were to work a forty-eight-hour week at rates of pay not less than 90 per cent of the rates paid for equivalent work in Montreal.[18]

The workers of Amherst celebrated their hard-fought victory, but the town was soon to suffer a series of defeats from which it would never recover. Amherst's One Big Union was unable to reverse the process of deindustrialization already under way in the town. During the 1920s, many of the town's main industries began to collapse because they had not kept pace with new technology and were unable to compete with central Canadian products. Amherst had a local depression in the 1920s, with its industrial work force shrinking to about seven hundred employees by mid-decade.[19]

❧

While workers in Amherst were fighting for their rights, developments were unfolding in western Canada that would again bring Roscoe into contact with the law. In March 1919, Roscoe had heard from Vancouver that many socialists were heading for a western labour conference in Calgary. It would be a historic meeting, organized by the Socialist Party leaders discouraged by the refusal of the Trades and Labor Congress to adopt a more aggressive course against the conduct of the war, government censorship, and the jailing of war opponents. The socialists felt that with the soldiers soon returning home capitalism would face its greatest trial. As the next stage in the struggle they wanted to build a class movement that, instead of organizing workers separately along craft lines, would include all workers in a new organization, the One Big Union.

The Calgary meeting laid the groundwork for the One Big Union, and it won quick acceptance. Many western unions were enthusiastic about joining it. But it was only weeks before the OBU's momentum was taken over by events out of its control. In Winnipeg, building and metal trades workers went on strike. Two weeks later, almost thirty thousand Winnipeg workers, both organized and unorganized, struck in support of their colleagues and on behalf of the principles of collective bargaining, better wages, and improved working conditions. Factories and stores closed and the trains were stopped. Within weeks workers across the West and other parts of Canada went out on sympathy strikes.

Influential businessmen, the media, and the government argued that the Winnipeg General Strike was a revolutionary conspiracy. The federal government refused to negotiate with the strikers. Instead, top strike leaders were arrested at gun point and taken to Stony Mountain penitentiary. The arrests took the momentum away from the strikers, and it appeared that a settlement was within sight. Before the strike ended, however, a group of Royal North-West Mounted Police, on horseback, charged into a group of strikebreakers on what became known as "Bloody Saturday." In the clash thirty strikers were injured and two died of gunshot wounds. The strike was called off four days later.

Of eight strike leaders charged with seditious conspiracy, four—W.A. Pritchard, R.B. Russell, R.J. Johns, and George Armstrong—were prominent members of the Socialist Party. William Ivens and John Queen worked for the *Western Labor News*, and A.A. Heaps and R.E. Bray were members of the Winnipeg strike committee. Police raided Socialist Party offices in Winnipeg and Vancouver, and among the materials seized were several letters that Roscoe and his socialist friends from New Brunswick had exchanged with the arrested men. When the charges were read in court, four New Brunswick socialists—Fillmore, White, Goudie, and Thompson—were among a long list of people named as co-conspirators. Legal action was not taken against Roscoe and the other Maritimers, but their letters became a small part of the huge volume of evidence used against the eight main strike leaders. In a letter to Roscoe just three months before the strike, R.B. Russell had warned him not to be provocative in his letters. He pointed out to Roscoe that an Alberta socialist had been arrested and charged over comments he had made about the federal government.

The Winnipeg strike trials began in late November 1919 and ran for four months. Coverage of the trials was front-page news across the country, and reports appeared frequently in *The Daily Gleaner*. Fortunately for Roscoe, who probably would have been in trouble with his employer, his name did not appear. Roscoe's correspondence with the socialist leaders was only a small part of the Crown's evidence in the case. One letter introduced as evidence was from Chris Stephenson to Roscoe, in which Stephenson said that *The Western Clarion*'s mailing list had been destroyed. The Crown hinted that the destruction had taken place with sinister intentions in mind. In his address to the jury, W.A. Pritchard pointed out that when the paper was banned it was also required by law to destroy everything in connection with the paper.

The Crown did not present evidence proving that any of the accused were trying to overthrow the government by force.[20] Instead it was clear that the strikers sought only the right to use collective bargaining to win a wage increase. Despite this, six of the leaders were convicted of conspiracy to overthrow the government. Russell got the stiffest sentence—two years—and the others got one year.

The brutal crushing of the Winnipeg strike should have been a signal to radicals such as Roscoe that the Canadian establishment would use every means within its power to prevent a working-class uprising. Even before the strike trials began, the Borden-Meighen government had introduced legislation in Parliament to turn the small, regional Royal North-West Mounted Police into the Royal Canadian Mounted Police, which would be nationwide with a huge budget of nearly $5 million and extraordinary powers to spy on any group or individual in the country. The RCMP could use Section 98 of the Criminal Code, which had been put into force during the strike, to arrest, detain, and intern people without trial, to raid premises, and to ban publications. This legislation remained in force until 1936, and Canada was the only allied country to retain oppressive wartime legislation during peacetime.[21]

❧

In 1921 a debate began within the socialist movement in Canada that dramatically changed the nature of Canadian socialism and ended up completely disrupting the Socialist Party. The Communist International (CI), based in Moscow, was emerging as a major force in world socialism, and some Canadian socialists were eager to belong to it and affiliate with other socialist movements around the world. In the Socialist Party many members began to balk at this idea when they learned about the International's conditions for acceptance: The Canadian movement would be substantially controlled from Moscow and the party would be known as the Communist Party of Canada.

Roscoe was strongly in favour of affiliation. He wrote in *The Western Clarion* in early 1921 that he believed the fear of being dictated to by Moscow was based on nationalism and racism. "'Dictation from Moscow' is a straw man, a bogey man," he wrote. In view of the oppression of socialism in Canada, he saw membership in the CI as a form of protection and salvation for the movement. He wrote that the Bolsheviks in Russia were the only workers in the world who were consciously and intelligently carrying on a class struggle, and he believed that Canadian socialists belonged in that movement.[22]

A referendum held among Socialist Party members opposed affiliation by a slight margin. Roscoe and several others left the party, which was badly damaged by the split and eventually disbanded in 1925. Many people also left two other socialist parties—the Socialist Party of North America and the Social Democratic Party of Canada—and joined the Communist Party, which was formed in May 1921 and officially known as the Workers' Party.[23] Roscoe was among the first to join the new party and received a small red Workers' Party card with the inscription: "No struggle too small. No struggle too great."

Supporters of the new party made a serious effort to establish a Communist movement in the Maritimes during 1922-23. Several party organizers from Toronto visited the region, and the Communist paper *The Worker* carried news reports from the Maritimes. J.B. McLachlan spoke in Halifax and Cape Breton in favour of the new party, and socialist politics dominated the 1922 United Mine Workers District 26 convention in Truro. In Cape Breton a large number of miners joined the Workers' Party, and in Halifax, where Joe Wallace was emerging as the socialists' spokesman, the Halifax Trades and Labor Council endorsed the Workers' Party.

Roscoe was working toward the establishment of a Workers' Party local in New Brunswick in early 1922. The RCMP intercepted a letter sent to Workers' Party headquarters from party business manager Trevor Maguire, who had been on tour in the Maritimes. Maguire stated that a party local would soon be set up in the Burton-Fredericton area to put the party in close contact with the farming community. But the Burton branch was never officially recognized by the Workers' Party. Unlike the Socialist Party, the Workers' Party did not have small locals scattered across the country, and the first official Workers' Party organization in the Maritimes, known as Maritime District No. 1, included all the Maritimes.

The Workers' Party—which would not be officially known as the Communist Party until 1924—was well on its way to being established across the country, but Roscoe's interest in the new political party was overtaken by other developments.

Elizabeth Fillmore (far right) and unidentified members of the family.

Roscoe and his father, Willard, 1915.

Selina, Roscoe's stepmother, and his father Willard, 1940s.

The Fillmores in Burton, 1923. L–R: Ruth, Roscoe, Dick, Alex, Margaret, and Rosa.

Roscoe, far left, in Siberia, 1923, with a crew of women crop harvesters.

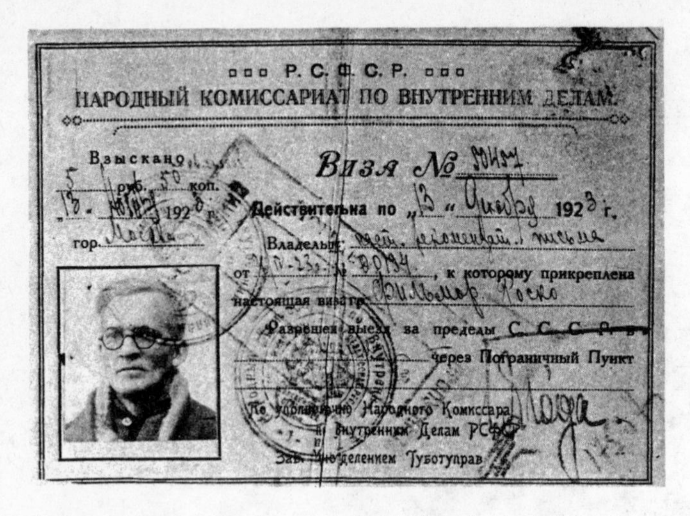

Р. С. Ф. С. Р.

НАРОДНЫЙ КОМИССАРИАТ ПО ВНУТРЕННИМ ДЕЛАМ

Взыскано 5 руб. 50 коп.

„18" 192 г.

гор.

Виза №

Действительна по „13" 1923 г.

Владельцу

от № , к которому прикреплена настоящая виза

Разрешен выезд за пределы С. С. С. Р.

через Пограничный Пункт

По уполномочию Народного Комиссара по Внутренним Делам РСФСР

Зав. Отделением Губотуправ

Soviet visa, 1922, when Roscoe worked in Siberia.

ELECT

The man we need
to protect our
interests

ROSCOE
FILLMORE

FARMER-LABOR CANDIDATE

Digby - Annapolis - Kings

—Full Employment—
—Policies to protect our farmers—
—Cold storage plant for Annapolis Valley—
—Rural and Urban Housing—
—Social Security & Higher Incomes—
—Health Insurance — Free Education—

Pamphlet for the 1945 federal election.

Granddaughter Lorraine outside the Fillmore home in Centreville, 1937.

The family home in Centreville, 1959.

A family gathering in 1944. Front, L–R: Frank, Nick, Lorraine, Irene, Becky, Back row: Rosa, Margaret, Flora Munroe, Bill Putnam, Roscoe, Dick.

Grandson Nick in the nursery station wagon, 1955.

Roscoe and Margaret, 1953.

Roscoe and grandson Ian, 1953.

Roscoe and daughter Allie in London, 1958.

Family members, L–R: Rosa, Roscoe, Margaret, Frank, Allie, and Dick, 1950.

Green Thumbs, *first published 1953.*

Roscoe speaking at Simpson's Garden Grove, in Halifax, 1955.

Roscoe's close friends Mabel and Charlie MacDonald in their cement home, next door to the nursery in Centreville, 1950.

At home in Centreville, 1961.

Looking over newly arrived nursery stock with a friend, 1960.

TEN

❦

A Journey to the "Promised Land"

EVER SINCE THE dramatic events of the Russian Revolution in 1917, Roscoe Fillmore had dreamed of going to see for himself what life was really like in the world's first socialist country. By early 1919 he had decided to see if he could make his dream come true.

Roscoe scanned newspapers and magazines sympathetic to the revolution for ideas on how he might get to Russia. Soon after the Society for Technical Aid to Soviet Russia was set up in the United States, he volunteered to go and work for them. He was not recruited, perhaps because so soon after the revolution Russia was in a state of crisis and did not place a particularly high value on horticultural skills. Several times a year he applied to the Canadian government for a passport to go to Russia. He was always turned down. Still hopeful, he began studying Russian, even though he didn't have much aptitude for the language. The events in Russia were all-consuming for him. When Margaret gave birth to their fourth child, a girl, in September 1921, she was named Alexandra Kollontai Fillmore, in honour of the Russian woman who fought for the emancipation of women during the revolution.

Relief groups and labour organizations were sending supplies to Russia, and Roscoe became a representative of one of these groups, the Friends of Soviet Russia in Canada. The Russians desperately needed tractors, tools, and supplies of all kinds to help expand their agricultural production. Joe Knight, one of the organizers behind the Friends and a member of the Communist Party, visited the Maritimes in the spring of 1922 and spent some time with Roscoe. Roscoe helped arrange meetings for Knight throughout the Maritimes.

When he realized his own efforts might not get him to Russia, Roscoe decided to seek the help of Big Bill Haywood, who was at that time living in Moscow. Haywood's self-imposed exile in the Soviet Union was the result of the U.S. establishment's efforts to crush the Industrial Workers of the World, which had experienced its greatest years of growth between 1914 and 1917.[1] More than a hundred thousand workers, including thousands of Canadians, belonged to the IWW, and Haywood had become the most feared and hated labour leader in the United States.

The U.S. government had tried many times to repress the IWW and finally saw its chance when the country entered the Great War in 1917. The IWW had condemned the war as a struggle among capitalists and said that working-class people should refuse to take part. The union faced cries of treason when it went on strike against several companies producing goods for the war effort. Government agents raided and occupied IWW offices, and Haywood and more than a hundred other leaders were arrested, jailed, and charged with "criminal syndicalism."[2] During a highly publicized trial in Chicago, little evidence of illegal activity was presented, but a jury engulfed by patriotic war hysteria convicted all of the defendants on every one of the ten thousand charges facing them.

Haywood was sentenced to the maximum twenty years in prison and fined $20,000. He spent two years behind bars before being released on bail awaiting the outcome of an appeal. He was fifty-one, in failing health due to diabetes, and shunned by the new, more conservative leaders in the IWW.[3] Afraid of losing his appeals and being forced to spend the rest of his life in a U.S. jail, Haywood decided, to the surprise of many IWW members, to skip bail. He boarded a boat heading for Europe and arrived in Moscow in May 1921.[4] Eight other IWW men convicted in the Chicago trials also took up residence in Moscow, most of them hoping, like Haywood, to make a contribution to the kind of society they believed in.

Roscoe knew from published interviews that Haywood lived in the Hotel Lux in Moscow, so he wrote him reminding him of their work together on the Maritime tour in 1909 and asking for advice about getting to Russia. Haywood quickly responded, recalling their experiences together and telling him that his qualifications as a horticulturist "would greatly enhance" his "value as a Russian citizen."[5]

Haywood had something specific in mind. Soon after he had arrived in Russia, Haywood had become involved in a plan to establish an industrial colony of U.S. workers in the Soviet Union. After three years of war with Germany and three years of civil war during which fourteen million people had died, the Russian economy was in ruins, and the government needed help to rebuild the country. Starting in the spring of 1920 there were almost two years of drought, and Russia was suffering from a devastating famine. When Haywood had arrived in the country, hundreds of thousands of people were dying of starvation. At home in Burton, Roscoe was agonizing over newspaper accounts from the Volga region that said more than twenty-five million people had no food. One report said, "In their terror, multitudes are fleeing—some to Siberia, some to Turkestan—but nowhere will they find a place of refuge which can absorb them and feed them. As their camps advance, cholera marches with them, and at the first touch of frost, if cholera ceases, typhus will begin."[6]

Aware of the desperate conditions, Haywood was one of many foreigners to become involved in industrial development and colonizing schemes to help the Soviet Union. He linked up with two other extraordinary socialists who had arrived in Moscow. Sebald Rutgers, a Dutchman, was a renowned engineer who became friends with Lenin. U.S. socialist Herbert Calvert, a member of the IWW, had worked in a Ford plant in Detroit.

Independent of each other, Rutgers and Calvert had prepared proposals that would employ highly skilled American, European, and Russian workers to develop and operate a major industrial complex at some site to be chosen in the Soviet Union. A top Soviet aide showed Calvert's proposal to Lenin, and the Soviet leader liked the idea. Lenin believed that the Soviet Union needed foreign know-how to rebuild the country. Together, Haywood, Rutgers, and Calvert devised a plan to establish an industrial colony that would help Russia alleviate its severe shortage of coal, one of the products that Allied countries refused to sell to the Soviet Union. The shortage was seriously harming the country's transportation and manufacturing sectors.

The group held meetings with top Soviet officials, and a location was chosen in western Siberia that had the ingredients needed to sustain an industrial complex. The Kuznetsk Basin, 2,400 miles west of Moscow, had several industrial sites that were either inoperative or not fully developed. There were coal mines and a steel mill operating far below capacity and a chemical plant that had never been completed.

By the time Haywood wrote to Roscoe in the fall of 1922, the proposal had been accepted by the Soviet government. The project was called the Autonomous Industrial Colony of Kuzbas—Kuzbas being short for Kuznetsk Basin. The first few colonists had arrived from the United States and were already working on various industrial projects. Little had been done in the way of growing grain or vegetables to feed the workers, but Haywood described the region's agricultural potential in glowing terms. "The soil is a rich, deep, black loam, of smooth texture," he wrote to Roscoe. "Abundant crops of all kinds grow; vegetable and garden truck is grown. Small fruits, such as raspberries, strawberries and currants grow in profusion."[7]

Near the mining town of Kemerovo, wrote Haywood, were a thousand acres of land well protected from the strong north winds by a two-hundred-foot bluff. "This strip of land I firmly believe will eventually be made an orchard and garden spot of the Colony. At present it is partly covered with brush and scrub trees, the clearing of the land will be comparatively easy." Haywood suggested that this seemed to be exactly the opportunity that Roscoe was looking for. He said that if Roscoe was interested in working at Kuzbas, he should contact the industrial colony's office in New York.

Roscoe wrote to New York and waited. In February 1923 he received a telegram that electrified him. Kuzbas wanted to hire about two dozen specialists, and Roscoe seemed perfectly suited for the position of head gardener. He was asked to go to Boston to meet a representative of the Soviet Union Council of Labor and Defence, the organization responsible for the Kuzbas project. Roscoe got ready to leave for Boston.

❧

In Boston Roscoe met with Harold Ware, an American representing the Soviet Council for Labor and Defence, and was offered a contract to become the head gardener at Kuzbas. Roscoe would supervise the production of crops for the 1923 season and then determine whether his whole family would move to Russia. Roscoe accepted the job without hesitation and hurried home to Burton.

Margaret was understandably worried about Roscoe's plan. Not sharing his strong political beliefs, she didn't feel the magnetic pull of the Soviet Union. She feared that Roscoe might fall into some unknown trouble halfway around the world, and it must have crossed her mind that she might never see him again. Then there were the four young children: Dick had just turned eleven and Ruth was nine, Rosa seven, and Alexandra just eighteen months. Margaret would have to look after this active brood all by herself. Haywood had told Roscoe that workers were not being allowed to take their children to Kuzbas at this time.

If Margaret had been the type to speak her mind, she would certainly have strongly opposed Roscoe's trip. Instead, she quietly expressed some concerns, and Roscoe allowed himself to interpret her response as a partial endorsement of his plan. According to Rosa, her mother's objections would not have been listened to anyway. "As far as Dad was concerned, it was cut and dried," Rosa recalled. "He was going. There was no question of anyone changing his mind."[8] When Willard was first told about the proposed trip, he "kicked like a steer." Nevertheless, when a group of about forty socialists and other friends held a going-away party for Roscoe in Moncton, Willard drove a horse and sleigh forty-five kilometres through deep snow from Albert to spend a few hours with his son before the departure. Roscoe believed his father was partially convinced that he would never see him again.

So he would have a job when he returned, Roscoe arranged for a leave of absence from the Saint John Valley Fruit and Land Company. But his relationship with his employer had become strained. Roscoe had been running the farm for nine years, he was growing tired of it, and he didn't like working for owner Arthur Slipp. Slipp, by then a county court judge, may have been pleased to see him leave.

It appeared that Roscoe's absence would not cause the family much financial hardship. The Soviet government made arrangements to send Margaret a hundred dollars on the first day of each month for the ten months her husband would be away. Margaret and the children planned to spend part of the time in Albert with Roscoe's family and part of the time in Springhill with her own family. The family left many possessions, including Roscoe's books, in the house in Burton, even though they weren't sure if Roscoe would be returning to his job at the farm. Friends and family, including Margaret's mother, made small loans to Margaret, and Roscoe would be paid enough so that he could save a small amount for his return.

The final step was taken when J.S. Woodsworth, the Labor MP from Winnipeg, helped Roscoe get a passport permitting him to travel to the parts of Germany not still occupied by French and Belgian troops. There was also a parting shot from *The Daily Gleaner*. When the paper heard of Roscoe's plans it ran a story stating that by going to work in Russia, "Roscoe A. Fillmore, advocate of Communistic ideas and active agent of Soviet Russia in New Brunswick as representative of the Friends of Soviet Russia in Canada, has decided to try 'a dose of his own medicine,' so to speak."[9]

Toward the end of March 1923 Roscoe caught the train to New York. At the Kuzbas office he signed the contract and took out a life insurance policy. All foreigners who went to work at Kuzbas also had to sign a statement that acknowledged the hardship in Russia and committed them "to work with maximum intensiveness and with productivity of labour and discipline on a higher level than capitalist standards, otherwise Russia will be unable to outstrip or ever catch up with capitalism."[10]

On April 5, Roscoe left New York on board the steamship *Hansa*, though a snow storm nearly prevented him from catching the ship. Unfortunately, his trunk didn't make it on board and was not located until many months later. Two men travelled with him and shared his cabin. W.H. Kingery, a middle-aged man who had sold and serviced farm machinery for International Harvester on the U.S. west coast, would become farm manager at Kemerovo.

Martin Zia, an expatriate Russian from Turkestan who had finished his education in electrical engineering in the United States, would be an interpreter for the two North Americans. In the hold of the *Hansa* was a shipment of farm machinery and equipment that Kingery was taking to Siberia.

For the most part, the ten-day voyage across the North Atlantic was pleasurable. The ship had band concerts and dancing. Recalling his strict religious upbringing, Roscoe wrote Margaret, "Imagine dancing on deck on Sunday afternoon!"[11] When other passengers on board discovered their plans, the three Russian-bound travellers soon became celebrities. Roscoe recalled that even first-class passengers showed enormous good will and curiosity about his hopes for Russia. But one old German woman issued a warning: "Buy lots of louse powder in Germany. You'll need it in Russia." Roscoe bought a pound of the powder, but never did find any use for it.

On April 14, the *Hansa* arrived at Cuxhaven, at the mouth of the Elbe River near Hamburg, Germany. One of the first things Roscoe saw, and something that excited him very much, was a huge, red Soviet flag flying on a ship that was going to Bremen to load cotton. In Hamburg Roscoe and Kingery attended a huge Communist Party meeting, although they couldn't understand much of what was said. Shopping in Germany was an adventure. "I bought a shirt yesterday," Roscoe wrote Margaret, "and when I got to the hotel and started to put it on, I found it as long as a night shirt! Well, the European shirts are all of a length. I shall have to cut a strip off this."[12] In every letter Roscoe raved about the wonderful but inexpensive food. After describing one meal in particular, he wrote, "And it was as fine a meal as you would get at the Royal Gardens in Saint John."[13]

❧

Germany's economy had not recovered from the war and was in disarray. When Roscoe arrived in the country, one dollar bought 21,500 marks. Ten days later, when he left, a dollar was worth 45,500 marks. Roscoe's U.S. dollars made German goods cheap for him, but he found that many working-class Germans were living in poverty. "The workers of Germany were living on potatoes and turnips," he wrote in his autobiography. "In fact, at that time, hundreds of thousands of workers were called 'turnip heads' because the allowance to the unemployed lost value so fast before they ever received it, that turnips were about the only diet they could afford."

Roscoe saw signs of the war's effect on the moral code of the country. What bothered him most was the drunkenness and the prostitution. The girls could not live on the stipend they received from the government. In the hotels in Hamburg and Berlin, young girls offered themselves as bedfellows for as little as a bar of soap. Roscoe wrote to Margaret: "Tonight in a restaurant one of the girls told Kingery, when he said he was cold, that he needed a hot water bottle with two ears. That's the damndest expression I ever heard in my life!"[14] Roscoe laughed at the time, but on reflection he felt dismayed by what he saw as a severe decline in morality. Despite this, he liked Germany. In Berlin he visited the graves of Rosa Luxemburg and Karl Liebknecht, the two heroes of the German socialist movement. "Like a damned fool I left my camera in Hamburg."[15]

Kingery stayed in Germany to look after the shipment of supplies, but Roscoe and Zia took the train on a route that would lead them through Poland, Lithuania, Latvia, and Estonia and on to Petrograd. In Lithuania there were a few tense hours when Roscoe and Zia were taken off the train and told their passports had been lost. They sat up all night in a tea room in a Lithuanian village, occasionally ordering a glass of tea and some bread so they wouldn't be kicked out. The next day their passports mysteriously reappeared and they continued on their journey.

Roscoe soon became homesick. He also felt guilty about leaving Margaret with all the family responsibilities. "You probably must have some idea how much I look forward to seeing Russia," he wrote. "That must be my excuse if any is needed for leaving you and the family for so long. But working as I have for something, and then having the opportunity to see it work out, was too much for me. Believe me, I shall miss you all damnably for the next few months, especially miss my dear baby Alex."[16] In every letter Roscoe wrote home during his trip he said how much he missed Alex.

Roscoe was a heavy smoker, and he was curious about the taste of Russian cigarettes. "I've just tried the Russian tobacco and it's the damndest dope I have ever tried yet! Full of wood, hay and horse manure, by the smell of it. Poor stuff. But I have several pounds [of tobacco] with me that I bought in New York. Have got it by the customs so far, guess I will get it into Russia alright."[17]

In Estonia, the last country before entering Russia, Roscoe saw the devastating effects of war. The worst destruction had occurred in one place where the war frontier was marked by a double barbed-wire fence that ran right through the middle of a town. On the German side was Eidkunen, but when the train passed through two gates, it was in Virballen. Roscoe and Zia had to stay in the Estonian town for the night. They were told they would be taken to a hotel. "And what a hotel," Roscoe wrote later. Travellers slept in a huge room where the walls were lined with several thousand rifles, lots of machine guns, and huge stacks of uniforms and helmets. It turned out they were sleeping in the town arsenal. Some of the roommates were villainous-looking types, which contributed to the effect. It was probably the only night Roscoe was scared in all the time he was away from Canada.

But Roscoe and Zia were not harmed and the next afternoon they took a train and travelled "hard"—a description for the poor conditions that existed in third class. They had no sleeping quarters, they were not allowed to eat in the dining car, and at night the only light in the coach was provided by smoky candles, lit here and there.

❧

On April 28, 1923, the train passed under an arch bearing the inscription "Workers of All Countries Unite!" printed in several languages. The travellers had crossed the border into Russia. For Roscoe, who was thirty-six, this was one of the greatest moments of his life. "I was deeply moved," he wrote later. "I had been a socialist for about 20 years...and I must have been affected somewhat as were the Israelites when they viewed 'the promised land' for the first time."

After crossing the border Roscoe continued to see the destruction left by the war. Railway buildings, telegraph poles, and bridges had been destroyed. There were wrecked buildings, miles of barbed wire, and forests of wooden crosses marking graves. The countryside was scattered with shell holes and trenches. Roscoe eagerly talked to dozens of Russians, with Zia doing the interpreting. A Russian who talked about the need for farm machinery in Soviet agriculture told Roscoe that the best land in the country produced about eight bushels of wheat per acre. Roscoe said that the same land in Canada would produce sixteen bushels per acre. He knew that Russia had a long way to go.

Roscoe was aware of strong differences of opinion about the success of the revolution. He kept files of newspapers clippings and was familiar in particular with a report based on a letter that Emma Goldman, the U.S. anarchist who was deported to Russia, had written during the summer of 1920 from Petrograd to journalist John Reed in New York. This was during the time of the famine, and Goldman had urged Reed to do everything he could to set up a relief organization in New York. She wrote that there was hunger in Petrograd. "The poverty and distress is beyond description. No medicines of any sort, no soap, matches, muslin, underwear and nothing of food. The suffering of the sick and of children is awful." She wrote that her own hands and feet were swollen from frost bite.

Despite the hardship, Goldman said the people still believed in the revolution. "In a temperature of 32 below zero, half naked and with King

Hunger ever present, the people go about their daily tasks in the passionate belief the revolution must be defended to the very last. They do a thousand extraordinary and amazing things that would test the vitality of people normally fed and clothed."[18]

Other people had less charitable things to say about Russia. Press criticism was divided into two categories. There was the blatantly anti-Russian press that carried lies and distortions to discredit the Bolsheviks. There were many stories saying that the Bolsheviks believed in free love. Some of the variations printed were comic. One story reported that Trotsky had eloped with the principal ballerina of the Moscow Opera and that, in their honour, Lenin had declared a love week.[19] But it seemed the most damaging criticism came from liberal British and American journalists who visited Russia. They argued that the revolution had lost its energy and that the country was controlled by a small group of people who restricted political and press freedoms—criticisms that Roscoe passed off as coming from people who didn't understand the difficulties the Bolshevik leaders faced.

Roscoe was relieved by what he saw out his train window. For a country ravaged by war, things looked quite normal. The people working in the countryside seemed healthy. There were no signs of anything like the famine Goldman had written about three years earlier. There was good food at the train stops, not just the basic staples of black bread and fish soup that he had expected to find. And he was not dismayed by the kinds of complaints he heard from passengers on the train. A well-dressed, elderly woman began talking to Zia and harshly criticized the Soviet government. She had been a property owner before the revolution, and she was angry that the government had seized her property. Roscoe wasn't sympathetic, but he thought it was a good sign that she and others openly criticized the government. He believed this was an indication that foreign journalists were lying when they wrote that people were jailed for criticizing the Bolsheviks.

When the train stopped at a junction, Roscoe got his first look at a Russian Orthodox priest. As people sat in a restaurant, "A theatrically gotten up person with long curly hair and a beard and clad in a long black gown entered. Not a person moved—not a person offered him a seat or even bothered to look at him. He posed in a supposedly Christ-like attitude of devotion and resignation for an hour or more."[20] Zia said that in the old days the people would have been crowding around to kiss the hem of the priest's robe. Roscoe saw this as another sign of the new times in Russia.

On the outskirts of Petrograd the train passed through flat marshes that stretched for miles, presenting a gloomy and monotonous landscape. Entering the city they saw more signs of destruction: huge buildings that had been levelled and factories along the bank of the Neva River that had been blown to pieces. Zia wept at the sight of the ruin. These were the same factories where the masses of starving workers who led the revolution had been employed. Roscoe looked upon the city and its people with reverence.

Roscoe arrived in the city with a letter of introduction to Commander Olga Lauki, a prominent figure in the Petrograd Soviet. She was the wife of Leo Lauki, one of the IWW members who along with Haywood had been handed a stiff sentence during the Chicago trials. Roscoe found Commander Lauki at the Astoria Hotel, which was the main office of the Petrograd Soviet government. That evening Roscoe attended a movie and dance in the ballroom of the Astoria as a guest of the Petrograd Soviet. He felt privileged to see the workers of Petrograd and their children enjoying themselves in a setting that had previously been restricted to wealthy tourists.

❧

Roscoe's greatest emotional experience in Russia was to witness May Day in Petrograd. He chose to attend the event in Petrograd instead of Moscow because of Petrograd's prominent role in the revolution and because it was much more distinctly working-class than Moscow. Everything Roscoe saw reinforced his faith in the revolution, and he wrote a description of the day's events for the *One Big Union Bulletin*:

> The day dawned bright and clear at 3 a.m. By 7 a.m. groups were to be seen from all directions making their way towards the headquarters of their various organizations. By 7:30 the crowds had increased and also companies of Red soldiers were to be seen everywhere. So complete was the tie-up of work—so important the holiday—that even at the International Hotel there was no attempt made to serve breakfast....
>
> At 9 a.m. I walked to the great square in front of the Winter Palace. The place was already crowded—masses of people everywhere. Boys had climbed into trees in the park and were singing The Internationale. A great reviewing stand had been built close to the wall of the palace. The buildings were gaily a-flutter with red flags and streamers and many large pictures of Marx, Lenin, Engels, Trotsky, Voldornsky, Uritsky (who was murdered by the counter-revolutionaries and after whom the square is now named), and many others.
>
> In the reviewing stand were Zinoviev, a number of Soviet official representatives of the various foreign consulates, the president of the Siberian Revolutionary Committee (who was the guest of Petrograd for that day), and a number of foreign newspaper men. All gathered in front of the former winter

home of Nicholas the Last and in the square in which, in 1905, the same "Little Father" had created a shambles for the workers on Bloody Sunday. 5,000 unarmed men, women and children were shot down—all had gathered to watch the workers of Petrograd make holiday.

And such a holiday! At about 9:30 came the vanguard from the northern part of the city. As they marched onto the bridge to cross the Neva they were saluted by the booming of guns from the Petrograd, Paul and Schrassberg fortresses. They came into Uritsky Square with banners flying, bands playing and thousands of men, women and children singing. Squads of Red soldiers were here and there in the line—Red sailors, Red artillery. They came in the thousands and tens of thousands. And the red banners they bore, typifying the common blood of the world's workers, came more and more frequently....

Further along the line, any friends [of] the enemy were represented by a bloated individual in a large cage transported on a motor truck. Another truck carried an enormous globe representing the earth. Maps were drawn on this, showing the outlines of the various countries. The whole globe was enveloped in a chain net. Only over the map of Russia were the chains broken, and a crew of workers stood on the truck and hammered continuously at the chains. Many similar scenes were depicted through the line of march. Printing presses were operated on trucks, turning out propaganda matter, proclamations, greetings, etc., and these were thrown to the people lining the streets.

The crowds were so great that they continuously encroached on the line of march. The militia, which is the police force of Russia, were employed in keeping the streets clear. And just here the contrast between Russia and America became evident to the writer. For in Red Petrograd, mark you, the whole purpose of the police was to ensure the workers every opportunity to parade and celebrate in perfect safety. Instead of battalions of burly, brainless, corrupt sluggers as in New York, the police force line is made up of class conscious workers. There was no rough stuff. The cry was continuously, "Please, comrades." The crowds good naturedly, singing all the time, broke up, pressed back and cleared the line. This happened hundreds of times.

Don't forget, comrades and fellow workers, that I am writing of a day spent among the "barbarous Bolshies," the "murderers" and "free lovers" of Russia. They crowded all about me, yet I didn't have my pockets picked nor my throat slit....

Noon came, three o'clock came and went, and still the workers' holiday and parade continued. None had stopped for dinner. They passed munching their black bread. And as they passed a battalion of Red soldiers they shouted: "Hail to the soldiers of the Workers' Revolution!" The soldiers replied: "Hail to the World's Workers!"

> And now came the Red Army—the pride of the workers of Russia and the class conscious workers of the world, 200,000 of them, foot, horse and artillery. Sailors of the Baltic Fleet and from Kronstadt, squads from the air fleet. New uniforms and equipment, fine upstanding young men and beautiful, spirited horses. It is easy to believe this the finest army in the world today. It is drawn from the people—the workers and peasants—the people who run and govern Russia—the people who have won everything from the Revolution and have everything to lose from its overthrow. And the Red Army is a university to these peasant boys. When they quit the army at the end of their service, they can read and write—they know the constitution and fundamental laws of Russia—they have been taught the fundamentals of communism and just what is the object of the Revolution. They go back to their villages in the far away corners of the Soviet Republic enthused and ready to work among their people towards the strengthening of the Workers' Soviets.[21]

This astounding day-long parade ended at five o'clock, leaving Roscoe convinced—if he ever had any real doubts—about the positive feeling of the workers toward their new government. He had watched for eight solid hours, and the final sight confirmed all the others: twelve large new street cars, "gaily decorated with green boughs, red flags and streamers" and loaded down with happy, singing children. He wrote, "The cars bore an inscription in gold leaf, showing they had been contributed and built in overtime and unpaid labour, by the workers of Petrograd and presented to the Soviet city on May Day."

That evening Roscoe walked up to Nevsky Prospekt, where the holiday atmosphere continued. He saw torch light processions, singers and musicians, and moving pictures of the parade projected on the outside of buildings. Later on, about eleven o'clock, he visited the Communist Club, housed in the palace that had once belonged to the Grand Duke Sergei.

> Here in marble halls, on beautiful tables of precious woods, under enormous chandeliers of cut glass, the former slaves of he and his class drank their tea and ate their bread and cakes and made holiday. I went up to the bridge and found the Neva a blaze of lights from boats which were loaded with singers, everywhere was song, laughter, joy. The prostitute American journalist who wrote that one never heard a happy laugh in Russia, nor yet saw a smiling child, wrote his stories from Warsaw, Copenhagen or Honolulu. He didn't see Russia. This is today [illegible] the happiest people in the world.

For Roscoe, the experience was "the fulfillment of a dream of almost 20 years."

> For almost 20 years I have been a member of various socialist organizations. For years I have posed—I think it was a pose—as something of a cynic. I have prided myself that I had cut out the sentimental or reduced it to a minimum—that I could as I fondly believed, view the whole situation philosophically and cold bloodedly. I who had prided myself on this, found myself laughing, singing and crying all day long. I have never in my life been so torn by emotionalism.[22]

Roscoe wrote Margaret that Petrograd was the first city he had seen in his life that he would like to live in. Many things he saw when walking around the city convinced him that, although there was hardship, many of the stories he had read in the North American press about Russia were untrue. He saw plenty of food in the shops. Zia said the peasants were dressed just as well, if not better, than twelve years before when he had left Russia. There was some prostitution, but much less than Roscoe had seen in the cities of Germany. Roscoe was impressed by the quality of the furs in the shops and stopped to buy fur collars for daughters Ruth and Rosa and a fur collar and cuffs for Margaret.

Curious about whether Russians were actively involved in religion, Roscoe visited four churches. They were all in good repair and providing services for those people who wanted to attend. In one of the Orthodox churches he visited during a service, there were only about a hundred people. The small attendance may have been due to the fact that the government discouraged religious activities. Roscoe was greatly impressed by the church. "It was the most beautiful place I ever was in my life, absolutely wonderfully decorated in gold mostly," he wrote to Margaret. "Everybody stands up, there are no seats. You stand or kneel. The priests sing the service and there is no denying that it is beautiful. The priests do not shave or have their hair cut. Their hair is about a foot long or thereabouts and really looks rather nice. Tomorrow we expect to visit the most famous church in Russia."[23]

Three days before he was to leave the city Roscoe met Olga Talanker, a young woman studying music in Petrograd. Olga was a revolutionary, and Roscoe was a socialist from a part of the world she knew very little about, so they had a great deal to talk about. Olga proved to be an ideal guide to the city, and together they went to a vaudeville show, saw the ballet, and viewed a Pushkin opera, *Eugene Onegin*, starring Leonid Sobinoff, whom Roscoe described as the Russian Caruso. In the few short days they became close friends. Years later, after Margaret had died, Roscoe wrote that he had one very brief affair during his forty-five years of marriage: He was probably referring to his meeting with Olga Talanker.

ELEVEN

Siberia: A Gardener for the Revolution

THE NEXT DESTINATION was Moscow, but once there Roscoe Fillmore and Martin Zia had only a four-hour stop between trains—just enough time to take a car ride to see the Bolshoi Theatre and Red Square, where John Reed, who died the previous year of typhoid, was buried. Roscoe didn't even have time to see Bill Haywood, who was expecting his visit.

W.H. Kingery caught up with Roscoe and Zia in Moscow, and the three companions boarded a train for a tedious four-day trip to Siberia. The first scenery they saw, on the outskirts of Moscow, was similar to what Roscoe was used to in Canada: rolling meadows and wooded stretches, dotted with white birch trees. They passed through Ekaterinburg, where the Czar and his family had been executed after the revolution. As the train approached the Ural Mountains the landscape turned to rolling hills covered with pine and fir forests. The train moved slowly, stopping at station after station, sometimes staying put for hours.

Once they were through the mountains and moving into Siberia, the countryside became flat, much like the Canadian prairies of Roscoe's grain harvesting days. They passed through the northern Volga Valley, the part of Russia that had been hit hardest by the famine, but saw no signs of famine now. Further along they did see signs of the bitter civil war that had followed the revolution. There were trenches, barbed wire, and burial grounds with thousands of small wooden crosses. The White forces' leader, Admiral Alexander Kolchak, had led his forces against the Bolsheviks in Siberia.

The train entered the Kuznetsk Basin on May 15, 1923. The scene was picturesque, with hills covered with birch and pine trees, and the Tom River dividing the town of Kemerovo. On one side of the river, where Roscoe would live and plant a garden, were the coal mines. On the other side was the chemical plant. The Kuznetsk Basin had all the necessary ingredients for a major industrial centre that could produce the materials and chemicals needed to develop the region. The goal of the Autonomous American Colony of Kuzbas was to use the power of the vast coal deposits, estimated at one-quarter of the world's known supply, to run a giant chemical plant partly completed before the war. It was the only chemical plant in the region.

The Kuzbas Colony was set up in an established town of more than ten thousand people. The first group of about seventy colonists had arrived at Kemerovo almost a year earlier, in June 1922. Many of them belonged to the Industrial Workers of the World, and they brought with them some romantic notions of the revolution—notions soon dispelled. They tended to dislike the traditional management techniques that had been put in place. Living conditions were dismal: The buildings that were supposed to house them were still under construction, so some of the colonists had to live in railroad boxcars; others lived in tents. Sanitary conditions were crude. There had been an outbreak of typhus, and a four-year-old American girl had died from dysentery.

In September 1922 twenty colonists had left Kuzbas, and another thirty-seven had left during the following winter. The Russians had talked about cancelling the project because it was costing them too much money. During the winter of 1922 colony manager Sebald Rutgers had gone to Moscow so that he and Haywood, who was working in the Kuzbas office there, could try to save the colony. Fortunately, they got support from Lenin, who saw Kuzbas as a means of tapping the technologies and experience of advanced capitalist countries. Rutgers was successful in reaching a new agreement according to which the Soviets would invest two million rubles in Kuzbas.

The Kuzbas people in New York were also having problems. They were trying to get the communist and liberal-left community to help them recruit skilled workers for the colony but were hampered by the colony's identification with the radical image of the IWW—an image now unpopular. Kuzbas was further damaged by the criticisms of the disgruntled colonists who had returned to the United States. Some former colonists helped launch politically motivated criminal charges against the New York officers of Kuzbas, alleging that the officers had committed theft by not returning money that belonged to the colonists. The charges were dismissed in court.

Despite the problems in Moscow and New York, conditions at the colony had improved by the time Roscoe, Kingery, and Zia arrived. They were the first new colonists since August 1922. With many of the colonists now lodged in the newly constructed Community House, adequate sanitary conditions had been established, and there was plenty of food. There were about 400 people at the colony, including 233 men, 80 women, and 80 children. Although it was called an "American" colony, only about 10 per cent of the people who came to Kuzbas from the United States had been born in that country. They came from all over the world and spoke many different languages. At least three other Canadians, in addition to Roscoe, spent time working at Kuzbas.

❧

Kingery quickly took charge of the huge agricultural lands, located about thirty kilometres from Kemerovo. The colony had been assigned twenty-five thousand acres to be developed into a model farm that would stimulate agricultural development throughout the region. Using the heavy U.S.-made tractors and other machinery the colony had acquired, Kingery began to plant crops of rye and other grains. The Siberian peasants and their crude tools could never have matched what Kingery quickly accomplished.

Kemerovo, midway between Moscow and Vladivostok on a latitude a little farther north than Edmonton, had a short but hot growing season during which daytime temperatures could reach 100 degrees. Roscoe's task was to grow several fields of garden vegetables to feed the colonists and the Siberian workers. The piece of land had never before been cultivated, but, as Haywood had said, the soil was fertile and the land was protected from the north wind by a steep bluff at the side of the Tom River. Roscoe began organizing workers to plough the new garden. Crews of women from the colony and the town soon planted thirty-five acres in different fruits and vegetables, including beans, peas, lettuce, tomatoes, cabbage, cauliflower, radishes, potatoes, and melons.

Meanwhile, progress was made at the Kuzbas industrial sites. During the first year the main coal mines and chemical plant had been under Russian control. There had been resentment and friction on both sides. The Russian managers believed that they could run the plants just as efficiently as the Americans, and the colonists were eager to prove them wrong. In February 1923 the colonists took control of most of the industrial sites, including the coal mines, chemical plant, sawmills, and machine shops. The Americans reduced the large bureaucracy the Russians had built up, and the U.S. mine manager was able to double productivity. By the end of the year Kuzbas employed more than two thousand foreigners and Russians, and the chemical plant was due to begin production in 1924.

Many of the residents of Kemerovo felt uneasy about the influx of foreigners, worrying especially about the colonists taking over their jobs. The situation was not helped by a tragedy that occurred within a week of Roscoe's arrival. There was no bridge across the Tom River, so foreign engineers built a cable system to guide a small boat that could transport passengers and supplies. Several uneventful crossings were made the morning the service was inaugurated, but in the afternoon, when the helmsman was not there, a group of impatient passengers decided to make the crossing on their own. Halfway across the boat capsized, tossing people into the swift current. Eighteen people drowned, all but one of them from Kemerovo.

Most of the local population were Tartars, and they lived under conditions much more severe than those faced by the colonists. Their housing was in bad repair, sanitary conditions were poor, and many lacked proper clothing to protect them from the bitterly cold Siberian winter. Both the birth rate and the mortality rate were high. Memories of the civil war were fresh. White and Red forces had controlled opposite sides of the river and many townspeople were killed or maimed in the ensuing battles.

More than forty Tartars worked for Roscoe, mainly planting and harvesting vegetables. They were paid either with meals or rations on a much lower scale than the colonists. The government's rationale was that the colonists would produce great benefits for the economy from their industrial work and therefore should get higher pay. The colonists tried to increase the pay of the Siberian workers but were blocked by the Russian trade unions, which feared the workers would be "spoiled."

By late June much of Roscoe's vegetable farm had been planted. He wrote Margaret saying he had two good foremen and didn't have to spend long hours in the fields like he had in the beginning. But there were problems. Some of the gardening equipment had not arrived, so they had to use rudimentary tools. One of the workers had nearly killed himself by accidentally driving a tractor through a bridge. Later in the year a former German war prisoner who had been running the greenhouse was accused of stealing produce and selling it at the market. Roscoe had to remove him until after the trial, replacing him with a man who had little experience, which in turn made more work for Roscoe.

Roscoe wrote home that he liked his living quarters, the small room he shared with Zia in the Community House. They had running water and electric lights, albeit unreliable ones. There was a community laundry that would clean ten pieces of clothing a week for each colonist. Ruth Kennell, a journalist who wrote articles for *The Nation* in New York, ran a small library. There was also a shoe repair shop, a school for children who couldn't speak

Russian, and a small newspaper. Everyone in the colony contributed 60 per cent of their earnings to pay for lodgings and food and health services.

While some colonists complained about a shortage of basic food supplies, such as sugar and butter, Roscoe was enthusiastic about the fare. "We had pancakes and syrup for supper a day or two ago," Roscoe wrote home. "They were good. We have porridge for breakfast regularly—oat meal, millet, Pearl Barley or rice. The millet is very good. Lots of milk here and cheap. Sunday one of the girls who works in the field brought me a jar of cream, sour cream. I guess you never tried it. I never did until I came to Europe. It's great stuff. You dip your bread in it. When it gets real sour you can spread it on your bread like butter. It's much better than cream cheese." But Roscoe ate very little meat. "I don't like the way it's cooked—always slopped into a gallon of gravy. Damn their gravies!"[1]

One of the few frustrations was a lack of news about what was going on in the rest of the world. Margaret mailed Roscoe packages of newspaper clippings but only one ever arrived. Roscoe assumed the material was stopped by Russian censors. He also believed that news received by wireless and posted at the Community House was censored. Whenever anyone received a batch of clippings or a complete newspaper, it was stuck up on the bulletin board for all to read.

There were few opportunities in Kemerovo for Roscoe to learn about how the socialist government in Moscow was running the vast country. He knew that economic conditions were still desperate, but he had no contact with government officials other than the people directly involved in running the colony. He was unaware that a struggle was building in Moscow to determine who would succeed Lenin, who had suffered his first stroke.

Roscoe's letters home reflected his excitement about the things going on around him. He enthusiastically supported benefit parties and sports events held to aid Russian military personnel stationed in the area. He wrote to Margaret: "Tonight there is a big spree. There will be an entertainment and dance here in the Community House for the benefit of the Red Navy. And twenty gallons of ice cream have been made, and I'm waiting patiently to get a chance at it.... We will buy big dishes tonight for five and ten rubles—that is about three to six cents. Also there will be cakes and candy. And I shall certainly enjoy watching the Russian women dance. They can dance beautifully. If I see half a chance I shall try to learn to dance. Everybody looks forward to a big time tonight."[2] There was a two-hundred-seat theatre in Kemerovo's People's House that had live theatre, put on by both the colony and the Russians, and silent moving pictures. Roscoe enjoyed the pictures but complained that it was too hot because they didn't ventilate the theatre.

Roscoe thought it was funny that everybody read the movie dialogue-titles aloud.

It seemed from his letters that Roscoe felt perfectly content in Siberia. "Remember me to all who inquire," he wrote home. "Tell them I am having the time of my life, that I wouldn't have missed this trip for the world. This is a great course of education. The experience is wonderful. I have wanted to see Russia for so long and felt that it was so difficult a proposition that now I seem to be in a dream."[3]

❧

After only a month at Kuzbas, Roscoe wrote Margaret saying he wanted to live in Russia. "I wish you would let me know by return mail whether you would be willing to consider coming over here. I myself favour it as Russia is pulling out of the hole fast. The hardships you have read about don't exist. We are fed well and well housed.... I expect our people, yours and mine, would raise hell, but after all, it is up to us. With all our scraps and misunderstandings, I think we care enough about each other and for the kids to be able to make a home for ourselves and be fairly well satisfied. If you and the children were here now, I would feel that I hadn't a care in the world."[4]

Margaret chose not to answer Roscoe's question about coming to Russia. Quite understandably she was afraid of moving four small children halfway around the world to a strange country that had just gone through a revolution. And no doubt some of the things Roscoe reported in his letters worried her. He said that from the small amount of news they got at Kuzbas, it looked like there was a possibility of war breaking out again. "If there is war, and I am needed, I shall go into the Red Army. This is the only country worth fighting for."[5] He said the workers were drilled regularly and held mock battles at night so they would be ready if attacked. He warned Margaret that if war came the Canadian or U.S. government might seize the hundred dollars a month she was getting from the Soviet Union. If the money was stopped, he said she should sell the car and the piano and raise what money she could from his insurance policy.

Near the end of June, Margaret took the four children to spend the summer with their families in Springhill and Albert. She seemed to be managing things well and there were no problems with the children, though Rosa later recalled that her mother was "a real lonely person" while Roscoe was away. Margaret wrote Roscoe at least once a week, often bringing him up to date on developments at home. She told him that the coal miners in Cape Breton had lost a bitter battle against the company and had gone back to work without an adequate wage increase. She said that Claude Davidson, a former member of the Socialist Party local in Albert who had just named his new baby Lenin, was interested in joining Roscoe in Siberia. The other news was that Fred Thompson, the New Brunswick socialist who had worked for Roscoe on the farm in Burton, had been jailed in California in April for organizing with the Industrial Workers of the World. Thompson was convicted of "criminal syndicalism" and was sentenced to four years in San Quentin penitentiary.[6]

Margaret also reported that Ruthie seemed to be getting negative comments about her father's politics from one of the preachers at her Sunday school. Roscoe sent back an angry reply. "I don't think little Ruth should be allowed to go to Sunday school and be tortured by those bastardly evangelists. God damn their stinking, lousy carcasses. I want her kept away from the poison dope!"[7] Margaret probably followed Roscoe's advice. When Roscoe wrote Ruthie he managed to get in a crack at the church without specifically mentioning her Sunday school problem: "Since I came here there has been a vote taken on the church question and everybody in this town voted to turn the church into a children's play and recreation house. Only five people voted against it in the whole place."[8]

Every letter that Roscoe sent home inquired about the children, especially Alex, the baby. He felt guilty for spending so long a period of time away from such a young child, and he was afraid Alex wouldn't remember him when he returned to New Brunswick. Roscoe wrote his son Dick a long letter, describing the progress of the garden, telling him about life at Kuzbas, and encouraging the boy to show an interest in Russia. "I think you should be over here boy," he wrote. "This is going to be the proper country for youngsters. As things improve in Russia the educational facilities will improve and that will be a fine chance for you to get a university education."[9]

By the end of July Roscoe's garden planting was completed, and the thousands of tiny vegetables and seedlings needed rainfall and several weeks of hot Siberian sun. Most of the garden workers went off on two-week paid vacations. Roscoe started to make plans for the construction of a greenhouse in the fall and had time to see more of the local area. He visited the bazaar, where he bought honey and white bread and shopped for gifts for the family. He practised Russian and swam in the Tom River.

Roscoe was amazed by the varieties of wild flowers growing along the river and in the countryside. In June he had written home: "The fields are now a blaze of colour. Even the wild peonies are now in bloom, and sweet peas. There are hundreds of different plants that are cultivated in the nurseries in America that are wild here. Delphinium, for instance, are now about two-and-a-half feet high and will soon be in bloom. The bush honeysuckle is yellow and pink and now in bloom; just the same bushes exactly are in our front yard in Burton, and the river bank is a mass of them growing wild."[10]

Roscoe had fallen in love with Russia. He wrote to Margaret again saying that he wanted to return to Siberia the following spring with her and the children. He wanted to bring them to Kemerovo for a couple of years and then move to another part of Russia with a better climate. Roscoe and Kingery had even picked out a location where they would build homes in Kemerovo. "The place is beautiful; on the new street about 300 yards from the Community House among the birches," he wrote to Margaret. "The ground is flat and a fine chance for a flower garden in the front. I will take a picture of the place tomorrow."[11]

He waited for Margaret to say whether she was in favour of making the trip. But she never did respond directly. When she wrote and asked whether the children should be enrolled in school in Albert for the fall of 1924, Roscoe said it wouldn't be necessary if they were going to come to Russia in the spring. He reminded her that he had not enjoyed his work in Burton and said he had no other job prospects in Canada. "I am strongly in favour of coming to Russia but I wouldn't coax you to," he wrote. "When I come back [to New Brunswick] I shall have a bunch of pictures and a lot of facts. A move of that kind is too big a thing, and the distance is too great to undertake unless you were perfectly assured that it was a good move and felt fairly sure that you would be content. But it's the only opportunity you'll ever have probably to see such a long stretch of the world, and the whole thing is damn well worth seeing."[12]

During the summer more colonists arrived at Kuzbas. Roscoe estimated that at least twenty-five different languages were being spoken in the colony. The last of a shipment of farm supplies—including five tractors—arrived from the United States, but Roscoe's trunk, containing much of his personal clothing, had still not arrived. Several months later, when he was back in Canada, he was notified that it had showed up in Germany.

Siberia was not without its hazards. Roscoe wrote home that he had broken some teeth and couldn't get them fixed by the colony's dentist. He said it would take a trip to Tomsk, a nearby city of about sixty thousand people. In another letter Roscoe was critical of a law that said all drinking water must be boiled. He sometimes drank unboiled water but soon regretted his actions. He and several other colonists developed severe cases of dysentery. Roscoe lost a lot of weight but soon recovered.

In mid-August Roscoe drove out to have a look at Kingery's farm at

Topka. "It's about twenty-five miles out there," Roscoe wrote to Margaret, "and the roads are pure hell and Kingery drives like a perfect fool—about like [Burton store owner] Jim Goan when he is drunk." On the way they visited a co-operative dairy in the village and saw butter being made and packed in barrels and casks. Kingery's farm had nine tractors and many other machines at work. Roscoe felt that the people were doing just fine. "They are busy now planting rye—as fast as they thresh this year's crop they are putting the seeds right back into the ground for next year.... Russia has a fine crop this year and can tell the world to go to hell!"[13]

❧

By early September the vegetable harvest was well under way, and workers were kept busy picking the vegetables and fruit, making sauerkraut, and pickling produce. At the height of the harvest Roscoe had about forty workers in the fields. "We now have sixteen barrels of kraut, twelve barrels of cucumbers and thirty-six barrels of beans packed. These barrels are the size of oil barrels and larger. Have now hauled 4,000 cabbage out of the fields. We have about 3,000 heads of cauliflower to cut and pick in October and about 8,000 more cabbage to make up. The beans are pretty well done; about six barrels more. Cucumbers; several more barrels; about twenty-five barrels of tomatoes to make up in pickles and preserves and several barrels of green melons."[14] In another letter Roscoe reported that he had worked all night in the kitchen making a barrel each of chow-chow and tomato preserves.

By late September the weather had begun to interfere with the harvest. Heavy rain damaged the rye crop at the farm. In the last week of September there was frost on the ground and the area was hit by a hail storm. Roscoe was rushing to harvest as many vegetables as possible.

Surveying the harvest, Roscoe felt that his work at Kuzbas had been successful, and he was more keen than ever to come back to Russia the following year. His enthusiasm intensified when Kingery announced plans to bring his own family to Russia. Another friend, Walter Lemon, had his wife with him, and Stanford Woomer, a machinist from Pittsburgh, sent for his sixteen-year-old son to join him. In mid-September Roscoe wrote a lengthy letter to Margaret, bringing her up to date on recent developments:

> Yesterday, Rutgers, the manager of the enterprise and a personal friend of Lenin, asked me to stay for the winter as Kingery is leaving for America in about ten days and they want me to take charge of the agricultural department here while he is gone. I would have a secretary, but I shall have to tell them "no,"

> as it would be too much for you to come over alone—altogether out of the question. At the same time, I would in a lot of ways, like to stay and see what the winter is really like and also whether I could really handle the job....
>
> I know that you never saw as beautiful a country in your life. I feel absolutely at home here. I love this country as I never did Canada, even when I was a kid and patriotic as hell. For slowly but surely, they are building up a country to belong absolutely to our class—the workers—and I would like to stay and help rather than go back to make a profit for somebody even if there are things lacking here that we have in Canada.[15]

By then Roscoe also knew that what he saw as his only chance at a job in Canada—the farm at Burton—was unavailable.

> Yesterday I received notice from Slipp that my services would not be required at Burton. I have felt fairly certain for the past year or two that this was coming. This is what any person may expect of the swine out there. I got that place on a producing basis. I did without my salary and worked for a low salary and now "we don't need you any more." I shall never be content again to work for any of the bastards, even if I could get a job, which is doubtful.

Slipp maintained that the quality of Roscoe's work had deteriorated—which Roscoe later admitted was probably true because of his growing dislike for the job and Slipp's supervision of the farm.

By the time Margaret received this letter, she and the family had moved back to Burton and taken up residence in a rented house not far from Slipp's farm. It appears that she incorrectly assumed that Roscoe would get his job back.

Roscoe was disappointed and a little perplexed that Margaret didn't indicate whether she was in favour of coming to Russia. By mid-October his work at Kuzbas was nearly completed. He was to leave Siberia on November 1. He was reluctant to go because of his strong belief in the revolution, but he was also eager to get home and see his family. He bought some souvenirs and gifts, including fur collars, an embroidered apron, and, for himself, a pair of knee-high boots with Russian embroidering on them.

On the way home Roscoe stopped in Moscow to see Big Bill Haywood at the Hotel Lux. They had dinner together and reminisced about Haywood's 1909 tour of the Maritimes, the old days of the IWW, and life in America. Roscoe thought that Haywood looked ill and like "a shell of a man." Haywood believed that the many months he had spent in prison in the United States was the cause of his sickness.

Haywood was unhappy in Moscow. He didn't fit into the new Russian scheme of things. The Russians were building their own political and industrial bureaucracy, and the IWW anti-organizational approach proved unacceptable to Moscow's new rulers. Haywood was no longer playing an important role in the Kuzbas project. When he became depressed he found solace in whisky and the old Wobbly associates who drifted into his hotel room. Haywood's main project toward the end of his life was an autobiography he was writing with the help of American communist ghost-writers.[16]

For the trip home, Roscoe took a route through central Europe and, a few days before Christmas, sailed from Liverpool aboard the vessel *The Canada*, bound for Halifax. As he travelled he must have worried about the future that awaited him back home. He believed that if he and his family didn't return to Russia, he had no prospect of employment. Many employers wouldn't hire him because of his radical political activities. He wished that he had paid more attention to developing his writing skills so he could pursue a career as a writer or a journalist. Possibly, he thought, he might get work with one of the radical political parties back in Canada. A worse fate, which he dreaded, was the prospect of going to work as a labourer on the Saint John waterfront. He continued to think that the best plan would be to return to Kuzbas in the spring.

❧

In early January 1924 Roscoe arrived back at Burton. In his absence Clarence Martin, Roscoe's second-in-command, had taken over management of the farm and moved into the company house with his family. The Fillmores took a rented house just down the road. Margaret's first impression was that Roscoe had lost a lot of weight. But the family was relieved to have him back. Roscoe couldn't stop talking about the wonders of Russia. For the first time in the history of the human race, he said, he had seen a successful government whose whole interest lay in the welfare of the common people.

Roscoe immediately began making plans to return to Kuzbas. He sent Thomas Reese, the U.S. representative on the Kuzbas management board, a report with his comments on the progress of the colony. Roscoe's main criticism was that many of the people sent over to Kuzbas from the United States were not well suited to the work and type of life. Reese agreed with Roscoe, writing back that it would be more beneficial to obtain labour on the world market. Reese wrote that he had observed that too many of the so-called radicals were merely meal-ticket hunters who didn't want to do real work in a constructive way.

Reese approved Roscoe's request that Kuzbas pay his expenses for a trip to Ottawa, where Roscoe wanted to get plant supplies for the coming season. He planned to take hardy fruit trees from the Canadian northwest. Reese said that Kuzbas would pay Roscoe's way back to Russia and home again, but there was no mention about Margaret and the children accompanying him.

Roscoe's and Margaret's families both objected to the idea of Roscoe taking Margaret and the four children to Russia. Roscoe's father had not changed his views on socialism, and family members didn't like the idea of the young children being raised in a country where English wasn't the main language. Margaret's mother was especially opposed to the scheme. Roscoe argued with them, pointing out the great opportunities that he thought existed in the Soviet Union, but he couldn't change their minds. Even though he didn't give up entirely on the idea of going back to Russia, everyone assumed the idea was dead.

During the summer and fall of 1924 Walter Lemon, who was working in the agriculture department at Kuzbas, wrote Roscoe to bring him up to date on developments at Kemerovo. The storage cellar that Roscoe had built kept in fine shape through the winter, with no sign of frost. The vegetable farming that Roscoe had started continued to be a success, although the weather in the summer of 1924 was colder and there was too much rain. The tomatoes would not ripen, and 80 per cent had to be picked green to get them harvested before the frost. Despite some problems, enough vegetables and potatoes were grown in the summer of 1924 to feed all five hundred residents of the colony, and the surplus was sold in village markets. "Conditions here are steadily improving," Lemon wrote. "One can see that the people are better clothed than last year." He said they had more food and their housing was being "steadily improved."[17]

Earlier, during the winter of 1924, there had been a huge celebration when the giant chemical plant finally went into production. Ruth Kennell described the event in *The Nation*: "The dramatic moment of the day had come when the heavy door was lifted and in a burst of music, with clapping of hands and waving of banners, the flaming coke poured forth."[18] Lemon wrote that the plant produced coke, gas, tar, benzol, and other byproducts. A complete industrial complex had been created—coal mines, coke production, the chemical factory, and supporting machine shops—realizing the goal of Kuzbas. The little band of foreigners—many of them idealists and long-time radicals like Roscoe—had made their contribution to the revolution. By the end of 1924, according to the reports, all of the industries at Kuzbas were thriving.

But even before the industrial complex was fully completed, the future of

Kuzbas was being determined in Moscow by the death of Lenin on January 21, 1924. His passing was greatly mourned at Kuzbas, not only because he was the father of the new nation but also because he had been the colony's strongest supporter. In 1925 new economic policies were adopted across the Soviet Union, and while many of the colonists opposed the changes, Kuzbas was integrated into the general economy of the country. No more skilled technicians or workers were recruited in the United States. The new Russian managers who took over Kuzbas were soon involved in a scandal and convicted of mismanagement of a state industry and misappropriation of funds. By this time many of the colonists had drifted away, some of them returning to North America, others going to live and work in other parts of the Soviet Union. By 1931 there were only twenty-five colonists still working in Kemerovo.[19]

The years 1923 to 1925 were remembered as the greatest and most productive by the foreigners who worked at Kuzbas. Their skills and technology produced results for the Soviet economy that may not have come about for years under traditional Russian technology. But while Kuzbas was an economic and industrial success, it was less successful as a social experiment. Many American-style socialists, especially the IWW people, did not find a common ground of understanding with the Russian communists. Of the six hundred people who came to Kuzbas from the United States, about two hundred had left unhappy. There was never any equality established between the colonists and the Tartar workers, and the colony itself was operated along traditional hierarchical lines.

For Roscoe, the opportunity to play any role—no matter how small—in supporting socialism in Russia was an honour. Years later he wrote that he looked upon Soviet Russia as "the most daring and important human society" ever established. "I shall be proud of my infinitesimal role in it so long as I draw breath and consider myself fortunate to have lived in the century in which this took place and proved itself a practical solution to the problem of poverty. This is the reason it is so feared and hated the world over. For to abolish poverty, exploitation of man by man must be abolished, and this means an end to the greed that has thrived in much of the world."

Many of Roscoe's friends at Kuzbas went their separate ways. Sebald Rutgers continued to show an interest in Kuzbas after it was turned over to Russian management, but poor health prevented him from playing any major role. Big Bill Haywood lived the rest of his life in the Soviet Union. When he died in 1928, half his ashes were buried in the Kremlin wall, and the other half were sent to Chicago to be placed near the graves of martyrs hanged after the Haymarket riot. Haywood had not finished the last important project of his life, the writing of his biography. Walter Lemon is believed to have died of smallpox at Kuzbas soon after writing his last letter to Roscoe. Ruth Kennell left Kuzbas, travelled about Russia for several weeks as a secretary to author Theodore Dreiser, and returned to the United States to write children's books based on her experiences in Russia. W.H. Kingery brought his wife and two children to Kuzbas, and they later left to join an agricultural group working in the Caucasus. After that he taught at an agricultural school in the Philippines.

In 1954 Roscoe was travelling through the western United States where Kingery had been born and wondered what had happened to him. In Seattle to visit the Botanical Gardens, Roscoe found Kingery's name in the telephone directory. "I debated with myself whether to call him for some time but finally decided against it," Roscoe later wrote. "Terror raged all over the United States at the time and anybody who had ever been in Russia, even on a visit, let alone for years, was suspect. I didn't care to call attention either to myself or him."

❧

During the first few weeks after his return home, Roscoe travelled the Maritimes speaking to audiences about his experiences in Russia. He returned to the familiar union halls of Cumberland County, spoke in Moncton at the city hall, and talked to the miners of Cape Breton and the farmers of the Annapolis Valley. Roscoe's speeches were frank, telling audiences both the bitter and the sweet about Russia.[20] "I had not spent ten months in heaven, neither had I spent them in hell," he later recalled in his memoirs. He said that he had watched a great people trying to dig their way out of the mud and that he sympathized and helped in every way that he could.

After Lenin's death in late January, Roscoe's trip to Cape Breton included a memorial meeting in Lenin's honour sponsored by the Workers' Party in Glace Bay. On its front page *The Maritime Labor Herald* of Glace Bay announced Roscoe as one of the speakers. The crowd of four hundred people in Glace Bay's Russell Theatre was made up mostly of coal miners, who had been on strike for three weeks. The miners enthusiastically applauded a recent Russian film on the leaders of the revolution, workers' demonstrations, and the Red army and fleet. Because of the strength of communism among the miners, Cape Breton was full of secret police agents, and local agents reported Roscoe's activities in Glace Bay to RCMP headquarters.

As the prospect of returning to Russia began to slip away, Roscoe knew he would have to look for work. He applied for a job with people he knew who owned a nursery in Ontario, was tentatively accepted, and then turned down. He had been asked to provide a list of references, and years later he learned that several people he offered as references had thought it necessary to point out his Bolshevik sympathies. Roscoe began to think that if he couldn't soon come up with a job, he'd have to try to start a small nursery himself.

TWELVE

New Life and Politics in the Annapolis Valley

In Russia Roscoe Fillmore had received a letter from W.J. Sim, a Scottish farmer living in Nova Scotia's Annapolis Valley. Sim and his partner, Frank Parry, were socialists who operated a farm at Northville, a rural village ninety kilometres west of Halifax, and they wanted Roscoe to come and visit them and speak to meetings in the Valley about his experiences in Russia.

Roscoe visited Sim and Parry during his tour of the Maritimes in the spring of 1924, and he took an immediate liking to the Annapolis Valley. The gentle sloping farmlands of the Valley stretched for a hundred kilometres between two small mountain ranges in the western part of Nova Scotia. The mountains helped provide an excellent climate for farming by sheltering the Valley from ocean fog and wind, giving the area more hours of sunshine and a longer growing season than most Maritime locations. Because of the fertile soil, the temperate climate, and the relative accessibility of the area by sea, the Valley had been one of the first areas in Canada to produce a surplus of agricultural products for export.

The Valley was particularly well suited to growing apples. By the time of Roscoe's visit orchards planted before the First World War were in peak production, and Nova Scotia led all Canadian provinces in apple production. It appeared to Roscoe as though all the apples the Valley could grow would sell year after year in the markets of Britain. There was room for expansion because less than 25 per cent of the available orchard land in the Valley had been planted.[1]

During his stay in the Valley Roscoe became good friends with Sim and

Parry and—while he hadn't given up the idea of returning to the Soviet Union—he began to consider the possibility of buying a small piece of land near his new friends and setting up his own nursery. Roscoe hoped he might be able to make his living—much like his father in Albert—raising young apple seedlings to maturity and selling them to farmers throughout the Annapolis Valley.[2] He discussed his idea with Sim and Parry, and together the three men launched a search for a suitable piece of land.

Roscoe soon found four acres of good land in Centreville, a village about four kilometres north of Kentville. His total assets at the time were a few hundred dollars, but he managed to make a down-payment and then arranged to rent a small house in the village where the family would live while their own house was being built. He returned to Burton, and no doubt Margaret was less anxious about the uncertainty of establishing a new home in Nova Scotia than about moving to Russia. In May 1924 the Fillmores left Burton for their new home.

Centreville was located practically in the shadow of the North Mountain. Land grants had been made in the area as early as 1759, and several of the families were descendants of Planter and Loyalist families.[3] Like most other small villages in Kings County, it was partly dependent on the apple industry. Most families operated small mixed farms and derived a part of their income from apples, whether from harvesting them, making the wooden barrels they were shipped in, or helping to run the railway that picked the fruit up in Centreville and neighbouring villages at harvest time. The small, dispersed village had a sawmill, a blacksmith shop, a cooper shop, apple and potato warehouses, a one-room school, a railway station, and a general store that sold everything from harness to bulk molasses.

The Fillmores settled into their rented house and every day that summer Roscoe and Dick, who was twelve, walked the half-mile to their new property, where they built a house of concrete blocks. Their land was on the main road through the village, next to the home of Charles and Mabel MacDonald, socialist friends of Jimmy Sim. MacDonald, who owned a company that made concrete products, gave the Fillmores credit for construction material, and Roscoe arranged more credit with several other small suppliers. The MacDonalds also helped on the construction of the new house, becoming close friends in the process.

Even as their new house was being completed, it became more apparent that Roscoe still wanted to go back to the Soviet Union. He wrote to Walter Lemon, who worked in the agriculture department at Kuzbas, and inquired about people who could help him join other foreign-run industrial projects in Russia. Lemon wrote back that Roscoe should contact the Herald

Commune, which was being organized in the United States and had been given the concession to take over an old estate of about 2,500 acres near Odessa. Lemon pointed out that the area was an ideal place for growing fruit and said that Russia badly needed good fruit growers.

It seems unlikely that Roscoe followed up on Lemon's suggestion because events in Centreville soon took over. In late August Margaret gave birth to a fifth child, a son who was named Frank Harris Fillmore in memory of Roscoe's brother who had died in 1919. A few weeks after Frank was born the house was partially completed and the family moved in. Roscoe's dream of a life working to build the socialist revolution of the Soviet Union evaporated. This was a great disappointment to him, but he didn't seem to be bitter toward the family members who opposed him on that score.

It would be another year before there would be any income from the nursery in Centreville, and Roscoe was in dire need of income. In the autumn he took a job picking apples. He also leased and operated a small orchard for a three-year term, and the sale of the apples from the orchard yielded a good profit in the first year.

To cash in on the prosperous times Roscoe planted two thousand apple seedlings, planning to sell them to farmers in two years. But as his trees grew he was shocked to see that the British apple export market was collapsing. By 1926 and 1927 Nova Scotia apple shippers couldn't get a reasonable price for their produce. Exports sales were less than half the level they had been when Roscoe planted his seedlings. Roscoe had hoped to sell the seedlings for at least $100 per hundred but ended up selling them at a loss, for about $10 per hundred.

Roscoe believed the market collapsed because Nova Scotia apple growers had never given the fruit trade in Britain a square deal. One of the problems, he thought, was that every year thousands of barrels of substandard windfall apples were being sent overseas as graded apples. Valley growers were also criticized for refusing to upgrade their orchards and for intimidating officials who graded the product for export.[4] As a result, apples from Ontario and British Columbia and from other countries were starting to do better on the British market. Roscoe's one bad experience with the apple seedlings caused him to lose faith in the apple industry for all time. After that he never grew more than a handful of apple seedlings in any given year.

For the family's second summer in Centreville Roscoe made plans for a small additional source of income. He hoped to grow, from seedlings, small crops of tomatoes, cabbage, cauliflower, and a few other plants. Roscoe and Dick built a lean-to greenhouse on the south side of the house. They germinated the seeds in the warmth of the attic and then moved the tiny plants to the near-completed greenhouse. In late spring some of the vegetables were planted for the Fillmores' own use while others were sold to neighbours. The money that came in for the bedding plants didn't make up for the revenue lost from the failure to sell the young apple trees, but the family survived.

❧

With the uncertainty of the apple business, the Fillmores decided to try to develop a different sort of nursery than originally planned. They would grow and sell vegetable seedlings, perennials, and annuals—including pansies, which soon became a specialty. As supply permitted, they also would sell ornamental shrubs and trees. There were only a few small nurseries in the province, so it seemed a good business to start. They called the company "The Valley Nurseries" and placed ads in the newspaper suggesting that instead of buying out-of-province nursery stock, Nova Scotians could buy at home for less and give a boost to the local economy.[5]

Roscoe soon discovered the reason for the limited Maritime nursery business: The climate of the region was the despair of horticulturalists. In winter much of the coldest weather came when there was no snow cover to help protect plants; and in summer there was often a prolonged hot, dry spell. To fight these conditions Roscoe set out to develop his own line of plants, carrying out extensive tests to determine which trees, shrubs, and annuals would be adaptable to the Nova Scotia climate. Over the years he combed the countryside looking for hardy plants that he could grow from cuttings. One of the best sources was the federal government's experimental station at Kentville, where there were several hardy evergreens and trees.[6]

Roscoe particularly loved the area of plant experimentation, but his lack of formal education led to difficulties. He couldn't understand a lot of the technical terminology in the few sophisticated horticultural books and magazines he was able to obtain. But he worked hard to educate himself in his new occupation, reading complicated texts over and over and memorizing the Latin names of hundreds of common plants. His work as a propagator gave new life to plants, such as shrubs and perennials, that had not previously been able to survive in their harsh environment.

Roscoe found common ground between his interest in plant propagation and his political ideology. He saw the same dialectical forces applying both to nature and to society—"continuous change with the emphasis on progress." This progress was "due to modifications and adaptations forced upon us by necessity." Both in his propagation experiments and his political struggles he worked to stimulate such "change with the emphasis on progress." Roscoe had read Darwin with great interest and defended the theory of evolution against creationists, arguing that religion clouded their objectivity and that only communists could be truly objective scientists.[7]

At home it was impossible to separate family life from nursery work. Everyone except Alex, who was just starting school, and baby Frank helped out. The greenhouse, where Roscoe spent much of his time, was located just a few feet from the back door of the house. In the winter the family made apple tree root grafts in the living room and stored them in the cellar. In the spring and fall, when she wasn't busy cooking or looking after the children, Margaret worked in the fields. Dick helped Roscoe with the heavy work around the nursery while Ruth and Rosa helped with the weeding and other tasks.

In the third year of operation, a problem in the propagation process threatened the very survival of the nursery. Seedlings of several kinds of plants began to look thin and were obviously unhealthy. Each year the problem got worse. Roscoe eventually learned that the trouble was caused by a fungus that grew along the surface of damp soil and lived on the tiny sprouts that emerged from the seeds. He tried several recommended recipes for getting rid of the fungus, but without success. The nursery was unable to fill many of its orders, and the family's income was seriously reduced. It was only after three years of poor crops that the answer was found in a light spraying of a formaldehyde solution.

These early years with the nursery were an extremely difficult period for the Fillmores. Some weeks there was barely enough food in the house to feed the five children. The kids wore whatever hand-me-downs and second-hand clothes they could. Rosa was surprised at how difficult everything seemed in Centreville compared to happier times in Burton. Eventually Roscoe and Margaret decided that Dick, in grade nine, would have to quit school and go to work. With no job opportunities close to home, he was sent to Massachusetts to work as a gardener with a long-time family friend, Dick Kidder. It was a devastating blow for the youngster, who loved school. As Roscoe's father had done, now Roscoe's son was leaving the Maritimes to earn his living in the United States.

To add to Roscoe's stress, his grandmother Elizabeth died in Albert in the fall of 1927, two weeks short of her ninety-fourth birthday. Roscoe didn't go to Albert for the funeral because there was not enough money to pay for the trip. He still had mixed feelings about his grandmother. He knew that she had loved him, but he had a difficult time feeling affection for her. Roscoe believed that his strict upbringing under Elizabeth's direction had caused him to become somewhat repressed as a person.

Willard's tree nursery in Albert was not doing well. The business had become dependent on sales to the United States, so when an embargo was placed on Canadian fruit tree imports in the 1920s because of possible diseases, the nursery suffered. Willard and Selina were left with practically nothing but a small piece of land and a house—which in the 1920s in rural New Brunswick were worth very little. Selina was sick with pernicious anaemia, an often fatal blood disorder that left her tired much of the time, and she wanted to be near the children. Somewhat discouraged, they sold everything for a few hundred dollars in 1928 and moved to Detroit, where they lived with Roscoe's half-sister Mabel. Willard, in his mid-sixties, worked as a gardener in summer and shovelled snow in winter. Clara, Roscoe's sister Ellida, and Margaret's mother also lived in Detroit, which had replaced Portland, Maine, as the main point of migration for the Fillmores to the United States.

❧

Roscoe blamed himself for the financial problems at the nursery. He felt he was not good at business and thought that his other interests—politics, plant experimentation, and reading—interfered with his work. He saw himself as a romantic and he believed that Margaret was the more practical of the two. Margaret would remark only half jokingly that Roscoe could run the world but couldn't look after his own affairs.

The difficult times didn't help Roscoe's and Margaret's relationship. By Roscoe's own admission, he became something of a domestic tyrant. "I had inherited or acquired a great deal of my grandmother's abruptness and irascible temper," Roscoe wrote in his autobiography, "and years of overwork and financial problems didn't improve me." When Roscoe became unduly angry Margaret tended not to speak her mind. Usually she would just leave the room. "I have often wondered why she remained with me during this period," Roscoe wrote later. "I suppose it was the children."

Margaret was often the sparkplug that kept the family going. She looked after the family's finances, and it often was her ingenuity that put the next meal on the table. She was always hiding away small amounts of money—in a jar or in a secret place in the bedroom—and bringing it out when it was most needed. She knew where to get the best prices on food and how to get the most out of a meal. When she had to buy clothes for herself or the children she searched out the least expensive items in the Eaton's mail-order catalogue. When there was no money to pay for Ruth and Rosa to travel to school in Kentville, Margaret would go to the Centreville station master Prescott Neville and arrange credit, or when the nursery's truck was still at

Kidston's Garage because the bill couldn't be paid, Margaret went to the garage and told them that Roscoe had to have the truck, promising that her husband would pay the bill as soon as he had the money.

Margaret got her enjoyment out of going to the movies—sometimes walking the eight kilometres to Kentville and back just to see a show. She wrote a movie script once and submitted it to a Hollywood studio, which didn't reply. She took part in community activities—especially so Roscoe's politics wouldn't isolate the family—and occasionally attended the Baptist Church. She had the children attend church: Ruth and Rosa often sang duets for the congregation.

Neither Margaret nor Roscoe was very good at enforcing discipline, and a boisterous atmosphere pervaded the house. Schoolmates of the Fillmore children thought that both Margaret and Roscoe were loving and caring parents. But the anger that Roscoe sometimes displayed frightened the children. Roscoe would lose his temper and curse a long string of H-words. Rosa, in particular, was bothered by her father's "blood and thunder." She later said she believed this worked to prevent her from becoming interested in radical politics.

Roscoe's politics created problems at school for Rosa and Dick, the children most bothered by the teasing and taunting of their school mates. "When any argument came up in the school yard, someone would say, 'Oh, your father's a communist! You people don't know anything. You shouldn't even be here.'"[8] Of the four oldest children, Ruth was the one who most admired Roscoe's political beliefs. She was fiercely proud of her father and once beat up a Centreville boy because he teased her about communism.

The villagers of Centreville had mixed feelings about Roscoe. Most residents were Protestant, many of them strong Baptists just like Roscoe's ancestors. The common opinion was that Roscoe was a "well-read" man whose politics were his own business, though a few people thought he was a kook for believing in radical politics. At least one person informed the RCMP about Roscoe's activities on a regular basis. Despite this the Fillmores got along with just about everyone. Roscoe became a member of the school board and gained a reputation for being kind and generous to people who were less fortunate.

But Roscoe inevitably stood out because of his belief in taking political action. He followed with great interest the controversial case of two Italian-American immigrant labourers, Nicola Sacco and Bartolomeo Vanzetti, who had been charged with the murder of a guard during the robbery of a shoe factory in Massachusetts. The two men had been convicted and sentenced to die in the summer of 1921. But the belief that the trial had been biased led to the formation of a large movement protesting their treatment. Critics argued that the men had been convicted because they were foreigners and opposed to many things that America stood for. Appeals and protests were carried on for six years, but a state-appointed commission reaffirmed the verdict and the two men were executed in the electric chair on August 23, 1927.

Roscoe was so angered by the executions that he went to the offices of the Kentville weekly newspaper, *The Advertiser*, to place an In Memoriam advertisement in their honour. The newspaper reluctantly printed Roscoe's ad, but only after consultations with a lawyer and a severe revision of the content. The In Memoriam, which Roscoe signed with his own name, said that the two Italian workers had been sent to their deaths because their ideas were not in agreement with those of the ruling class of the state of Massachusetts. Roscoe wrote that while he did not endorse Sacco and Vanzetti's views on anarchism, "I glory in the fact that in times when everything is for sale... these two working men held solidly to their faith in the new social order."[9]

The advertisement caused an uproar. *The Advertiser* was so besieged by complaints from angry readers that the editor felt compelled to address the issue in two editorials the following week, one defending the fairness of the trial and the other stating that the paper had no sympathy with the views expressed in the ad and that anyone entertaining similar views would have to find some other medium of publicity. According to the editorial, in Russia a critic like Fillmore would surely be killed more swiftly than Sacco and Vanzetti. It concluded, "We in Nova Scotia can scarcely be blamed if we find it desirable to suppress opinion which is directed at the foundations of our society and institutions, in which we believe the happiness of ourselves and our children is established."[10]

On the Labour Day weekend just after the Sacco and Vanzetti storm, despite a small ad for the nursery placed in *The Advertiser*, there were very few customers—confirming one of Margaret's worst fears: that Roscoe's political activities would lead to a boycott of the nursery. Bertha Leslie, a widow from nearby Woodville, had seen Roscoe's In Memoriam ad in the paper and thought there might be a boycott, so she and her two sons, Robert and Ken, came and showed their support for Roscoe's right to his opinion by buying a large number of plants. Roscoe never forgot the gesture. As a person who supported unpopular causes, he often felt lonely. "Practically every day," he wrote in his memoirs, "men and women 'stick out their necks' in the interests of justice, fair play and freedom of expression, and in many cases find themselves practically alone, though the causes they fight will benefit the whole country. It would make a lot of difference if they were sure of the support and sympathy of a few others, even a baker's dozen."

❧

Beginning in the late 1920s and continuing well into the 1940s Roscoe was a key part of a small but lively network of free-thinkers and socialists in the Annapolis Valley—the only such group in Nova Scotia outside of Halifax and Cape Breton. The other core members of the group were James Sim and Charles MacDonald. The group often got together at the Sim farm in Northville, on the sloping hills at the foot of the North Mountain overlooking the apple orchards of the Annapolis Valley. The farmhouse provided the perfect environment for the group that became known as the Centreville Socialists. The gatherings, usually held on Sunday afternoons, were often aided by jugs of Sim's homemade apple cider.

The host, Jimmy Sim, Roscoe's closest personal friend, was a strongly built, sincere man who was well liked by all his friends, not least because he seemed to favour talking and debating politics over the hard work of farming. He became a socialist soon after migrating to British Columbia from Scotland in 1901. Sim and his friend, Welshman Frank Parry, had started a fish hatchery, and they both belonged to the Socialist Party, supported it financially, and subscribed to *The Western Clarion*. In 1912 the two men had moved to Nova Scotia, where they bought their land at Northville and set up a fruit farm. After he settled in Nova Scotia, Sim married Annabelle, his teenage sweetheart from Scotland.

Charlie MacDonald was a blunt, outspoken man of firm conviction who, at just over five feet, was dwarfed by his tall wife, Mabel. "Mac," as he was called, had subscribed to *The Western Clarion* for several years, had recruited new subscribers, and had occasionally published articles in the paper. He had become a socialist early in life when he travelled the world as a ship's carpenter on a clipper ship. Like Sim, he had lived in British Columbia, where he had worked on railway construction and built trails to the gold mines of the interior. In 1912 MacDonald returned to settle in Centreville, where he established his small concrete company. He applied many of his socialist beliefs to the operation of his company, paying workers nearly double the going wage rate and devising a plan to put aside a portion of employees' wages to be drawn on through the winter when contracts were scarce.[11]

A frequent guest at the Sim farm gatherings was Kenneth Leslie, an accomplished poet—one of the young men who had come to the nursery with his mother the weekend after the Sacco-Vanzetti executions. Leslie and his first of four wives—Elizabeth, the daughter of wealthy Halifax candy manufacturer James Moir—worked a farm a few miles from Centreville.

Boisterous and always with an amusing story to tell or a practical joke to play, Leslie was a powerful Gaelic singer and an accomplished violinist. He was strongly influenced by the social gospel, which made for interesting discussions when he got together with Roscoe and the other revolutionary socialists. Leslie's poems won acclaim in London and New York, and in 1938 he won the Governor-General's award for poetry for publication of *By Stubborn Stars and Other Poems*. In the early 1930s Leslie moved to Boston and began publishing a left-wing monthly magazine, *The Protestant Digest*, which played a prominent role in the fight against fascism and anti-Semitism in the United States.[12]

Other guests included Otto and Asta Antoft, a Danish couple recently arrived in Nova Scotia who were involved in publishing the nationally circulated newspaper *The Danish Herald* at nearby Lakeville, Nova Scotia. Also among the guests were several prominent members of the Communist Party, including party leader Tim Buck, *Canadian Tribune* manager Annie Buller, and party worker Beckie Buhay.

The meetings had no formal structure and it was usually only the men who took part in the political discussions. They often sat around the kitchen table while most of the women would get together in the living room. Everyone would come together when it was time to eat. Sometimes they'd sing old labour and socialist songs set to the tune of popular hymns. If, as happened on occasion, an RCMP patrol drove past the house, the officers wouldn't be able to hear the revolutionary words of the songs and might think that religious hymns were being sung.

Of all the people who regularly met, Roscoe was the one who most interested the RCMP. He was frequently in contact with Communist Party headquarters in Toronto, received party books and literature, and subscribed to *The Worker*. Roscoe, however, did not openly promote the Communist Party as he had the Socialist Party in earlier years. Of the estimated sixty paid-up Communist Party members in the Maritimes in 1925, probably only a half-dozen lived in the Valley.

❧

In the late 1920s, Roscoe, J.B. McLachlan, and Joe Wallace of Halifax were the most prominent Maritimers in the Communist Party. Even though Roscoe had only a tiny base of support in the Annapolis Valley, he was recognized as a pioneer in the Canadian socialist movement.

McLachlan was the most influential member of the small Maritime movement, largely because of his influence on the radical miners and steelworkers of Cape Breton. The island held much promise for the party, which sent a string of Toronto workers there. McLachlan also was well known because of his dramatic battles with heavyweight John L. Lewis of the United Mine Workers (UMW). In 1923 McLachlan enraged Lewis by trying to affiliate the Cape Breton miners with the Red International of Labor Unions (RILU). McLachlan was the UMW's district secretary-treasurer, and, a year later, Lewis deposed the Maritimer and his fellow officers and barred them from the union after they had led an illegal sympathy strike in support of workers in the Sydney steel mill. Shortly after, McLachlan was convicted of seditious libel for his role in the strike and spent four months in Dorchester Penitentiary. Barred from the mines, McLachlan ran a small farm to earn a living but remained influential in the labour movement by editing, first, *The Maritime Labour Herald* and, later, *The Nova Scotia Miner*.

Joe Wallace played an important role in the early development of the communist movement in Halifax, in spite of the fact that he was the manager of his brother's advertising agency. Strongly influenced by the labour radicalism of 1919, he became a Marxist, joined the Independent Labour Party, and was made editor of the Halifax labour weekly, *The Citizen*.[13] Wallace contested elections at the municipal, provincial, and federal level. When the Workers' Party was formed, Wallace—who would later gain recognition as a poet—helped lead a small group of radicals into the new party.

As leading Maritime communists, Wallace and McLachlan attended the Communist Party's national conventions in Toronto, and in May 1929 Roscoe was asked to attend. It was the busiest part of the year at the nursery, but Roscoe thought it wouldn't hurt if he were away for a week. He felt it was important that the area be represented at the national level, even though he knew the party had no hope of immediately making any gains in the conservative Annapolis Valley. Roscoe was not an official Maritime delegate, but was given voting privileges in recognition of his status as a veteran of the socialist movement.[14]

There were deep problems in the party, ranging from the bitter debates over new Soviet leader Joseph Stalin's attacks on Leon Trotsky to a disagreement about whether to accept the dictates laid down by the Communist International (CI). When the CI condemned "Trotskyism," it expected Communist parties around the world to follow its lead. It also condemned the existing socialist parties, such as the Socialist Party of Canada, believing they were hopelessly reformist rather than revolutionary. The CI declared that social democrats, such as J.S. Woodsworth, were dangerous elements in the ranks of the working class and ordered the party to have nothing to do with them.

The Canadian leaders were opposed to a new CI policy developed by Stalin that would have them abandon Lenin's policy of working within established trade unions in favour of building separate Communist unions.[15] Later, as part of the plan to reorganize the party structure, the CI also insisted that the Canadian party liquidate its foreign-language branches. This infuriated groups such as the Ukrainians and the Finns, who had made a major contribution to the party.

As it turned out, the convention Roscoe attended was one of the most crucial in the history of the Canadian party. The meetings became a battle for control of the party. Two slates of candidates were put to the convention. One was headed by party secretary Jack MacDonald, who, although he had sympathized with the views of Trotsky, was offering a compromise slate that included Tim Buck, director of the party's trade union movement and head of the pro-Stalin group. Roscoe voted for the winning MacDonald slate.[16] After Roscoe and the other delegates had returned home, however, it became clear that MacDonald didn't have the support of the Communist International and, during a series of bitter meetings, MacDonald resigned and Buck emerged as party leader. The Canadian Communist Party swung behind the Stalin leadership and never again in Stalin's lifetime wavered from that position.[17]

Roscoe was not pleased by what he had seen at the convention and came away somewhat disillusioned by what had taken place. He was an old-style socialist who had learned his politics long before the First World War, and he wanted a radical, united party that would be more interested in defeating capitalism than bickering about its relationship with Moscow.[18] After the 1929 convention he became less active in supporting the party.

THIRTEEN

The Dirty Thirties: The Poor Shall Want

By the summer of 1931 the Great Depression was taking its toll across Canada. World prices for grain had collapsed, and in the West thousands of men roamed from city to city looking for jobs that didn't exist. Unemployment in the country had already reached about 13 per cent and would soar to nearly 30 per cent by 1933. In Nova Scotia, there were six thousand unemployed union tradesmen in Halifax alone. In Cape Breton, more than seven thousand coal miners could barely feed their families because they were working only two days a week.[1] Business and government insisted that conditions would soon improve and did little to assist the increasing army of unemployed. Roscoe noted that even though the federal government had a surplus of $47 million it still rejected demands to introduce an unemployment insurance system.

Roscoe Fillmore was not active in the Communist Party, but he watched with interest as the party began rebuilding itself by organizing the hundreds of thousands of unemployed workers. But the party soon found itself on the receiving end of a Red Scare crackdown once again—this time a much stronger crackdown than in 1919. The police and other anti-Communist forces made full use of Section 98 of the Criminal Code, which gave them unlimited powers to arrest, detain, and deport people. The police attacked meetings of the unemployed and between 1931 and 1934 more than ten thousand people from across the country who were out of work were deported without an opportunity to properly defend themselves.[2]

On the night of August 11, 1931, police in Toronto arrested eight top leaders of the party, including Tim Buck. The party was outlawed. In highly publicized court trials the government tried to prove that the Communist Party was a violent revolutionary conspiracy. The prosecution's main witness was a labour spy named Jack Leopold, who had been discovered and expelled from the party in 1928. Even though the Crown's case was weak, all the defendants were convicted and sentenced to terms of three to five years.[3] Once again a group of radicals were jailed more for their beliefs than any proof that they had conspired to overthrow the government.

Soon after the 1931 Toronto raid, the RCMP again began to keep a closer eye on Roscoe, who still considered himself a communist. The head of the Kentville Detachment of the RCMP had become particularly interested in 1932 when he was called to the Fillmore home in Centreville to arrest a wandering drunk and by chance saw a framed picture of Lenin. Thereafter the Kentville Mounties regularly prepared reports on Roscoe's activities, and the officer on duty said in his report that he would attempt to determine the names of any persons known to associate with Fillmore.[4] Soon Roscoe's neighbour Charlie MacDonald and his close friend Jimmy Sim were being watched.

The police, ever on the lookout, confiscated periodicals and books that Roscoe received from the Soviet Union. Hundreds of books were forbidden entry into Canada, but nobody would provide Roscoe with a list of exactly what was prohibited. The Canadian Post Office refused to issue him money orders for papers or magazines published in the Soviet Union. The hypocrisy angered Roscoe. "Each year thousands of tons of the so-called 'pulp press' come in from the U.S.A., and its foulness stinks to the heavens," he wrote in *The Steelworker*, a radical weekly paper published in Cape Breton. "Royal Bank Bennett and his successor Mackenzie King are so anxious over the welfare of our Canadian workers that they cannot allow us to know just what the Soviet press says. Better that our people develop a sex complex than a class complex!"[5]

❧

The Fillmores themselves were still struggling to earn a living as the Depression hit. The apple industry was booming again, but because the Valley Nursery was only marginally involved with the industry, the business did not benefit directly from the success of the farmers. Like other families they did whatever they could to bring in money from new sources. They tried to start a miniature golf course across the road from the nursery but soon abandoned the idea after a season of poor weather. During the winters, when there was little income from the nursery, Roscoe earned extra money by managing the Centreville rink. Roscoe and other family members would still work together in the house to graft a desirable species of apple tree onto a hardy root stock. In early spring the root grafts were delivered to dozens of farmers throughout the Annapolis Valley who, in turn, planted them as part of their orchards—although many of them couldn't pay.[6]

Roscoe arranged with a store in Kentville to sell nursery stock on a consignment basis, and the Fillmores prepared cut flowers for occasions such as weddings and funerals. On Saturdays during the summer Ruth and Rosa, now teenagers, went door-to-door in Kentville selling cut flowers.

Roscoe also started to sell nursery stock at the City Market in Halifax. Every Friday morning before dawn he would load up a battered, half-ton truck and drive eighty kilometres to the city, where he would spend the weekend selling all sorts of plants at the market. Located on Brunswick Street just across from the Town Clock, the market was a centre of activity for many out-of-city green farmers during the Depression. Roscoe set up a table next to Hope McPhee, who operated a small farm at West Gore, Hants County. McPhee noticed that Roscoe wasn't selling very much nursery stock, but he enjoyed talking with him about politics.

Throughout the bleak period of the Depression, when those in authority were showing little concern for the thousands of people who were barely getting enough to eat, Roscoe played the role of social critic and agitator whenever the opportunity presented itself. He was particularly outraged by the injustice he found in Halifax. At the end of 1933, one-sixth of the city was living on relief.[7] Upstanding men and women, who had worked all their lives, were thrown into unemployment. It was a devastating experience, especially because there were no government programs to help out: no unemployment insurance, no family allowances, no organized welfare system, no medicare. People had little to fall back on other than the help of family members, unreliable private charities, and municipal relief, which was inadequate and administered in a way to humiliate people to keep them from applying.[8] Roscoe's response to the situation was to launch an attack on the callousness of the city's wealthy establishment.

> Halifax, a city of churches, of well-fed parsons, lawyers, doctors, business men and politicians; happy hunting ground of numberless shysters; allows 25-cents per week relief for the children of the workers. This means that milk, the most necessary and really only essential food for children, is out of the question, for milk is 12-cents a quart. Lacking milk, these children are mal-nourished; their bodies are starved; regardless of what other food may be available. And on 25-cents per week only bread and molasses can be given, and damned little of that.... This is a country that exports millions of pounds of butter, cream,

> cheese and milk; where thousands of tons of butter and cheese are always in storage; where over a half-billion bushels of various food grains—wheat, rye, barley, oats, buckwheat, beans, peas, etc.—are in storage at this moment, and government and private concerns racking their brains to find a market for these goods.... It is our class that is affected by this organized campaign of starvation against workers' children. The authorities will not see that they are fed unless we compel their attention by organization and demonstration.[9]

During his stays in Halifax Roscoe got to spend time with other communists who were protesting the lack of action being taken to ease the impact of the Depression. The leader of the party group in Halifax was Kingsley Brown, a boisterous and outspoken journalist who worked at *The Chronicle*. Roscoe attended meetings and social gatherings held at Brown's house on the corner of Robie and Cunard Streets. Around 1933 Brown was succeeded as Halifax party leader by Ralph Marvin, who worked at T.C. Allen's Book Store and was later a broadcaster on the CBC. Marvin's wife Ruth, the social editor at *The Chronicle*, was also a dedicated socialist. Marvin gave up the leadership when he feared he might be exposed as a communist.[10] He was succeeded by Dane Parker, who had returned to Halifax in 1932 after spending the first years of the Depression roaming North America in search of work. Parker, a small, nattily dressed man with a sardonic sense of humour, wanted to become a poet and a writer. He took an instant liking to Roscoe, who was older and had more experience as a socialist. The two men became lifelong friends. "I looked upon Roscoe as one would look upon their own father," Parker said.[11]

When Roscoe met Joe Wallace, the life of the prominent communist was in a state of chaos. Outraged by Wallace's radical views, some Halifax businesses had boycotted the advertising agency where he worked. Given the choice of either leaving the job or dropping his political activities, Wallace quit his job and went on relief. Giving away all his possessions and choosing to live in poverty, he said he was going to devote his life to socialism and poetry. He began to publish a newspaper about the problems of the unemployed, *The Hunger Fighter*, but it lasted only a few issues. In 1933 he moved to Montreal, where he began to work for the Canadian Labor Defence League.[12]

Among those who helped the party was Bill Ross, a hard-drinking, fun-loving organizer of the unemployed. Jailed at least three times for his work on behalf of the Canadian Labor Defence League, Ross had written a folk song, "Ballad of the North Sydney Jail." He owned a house on the Northwest Arm in Halifax, and on weekends many of the party faithful would take a small ferry across to the house for a social get-together. Another party supporter was Svend Gaum, a dry and droll Swede who worked at odd jobs.

The few people who worked as organizers for the party were paid practically nothing. They seldom had enough money to buy food and clothing, and they lived in the cheapest rooms they could find. Eddie Sarman, who headed the small Halifax branch of the Young Communist League, sometimes received a payment of three dollars from the party in Toronto. He drank bitter tea to fight back the hunger. Bill Findlay, a politically ambitious Scot, ran the party organization in Halifax in the mid-1930s and would spend hours in his small rented room preparing reports for the "centre" in Toronto.

One of the few communists who had a regular income was Conrad Sauras, a generous but hot-tempered cook from Spain who owned a working-class restaurant across from the Nova Scotian Hotel. Pencil-thin and loaded with nervous energy, Sauras put his radical political beliefs ahead of his business, often spending the restaurant's profits to provide free meals for the victims of the Depression. During the Spanish Civil War, the waitresses at the restaurant wore Mackenzie-Papineau badges in honour of the Canadians fighting fascism in Spain.[13]

As their concerns about the Depression grew, the activists got in the habit of meeting in the evenings on the Halifax Commons, behind Citadel Hill. On Friday night they would go to the corner of the Commons, across the street from the Armory, and wait for the ball game to finish. Then one of the socialists—often Roscoe—would jump up on a bench or a chair borrowed from the Armory and begin to address the people who had been watching the game. Roscoe was in his mid-forties, and his skills as a speaker had greatly improved from the earlier days when he had delivered angry speeches for the Socialist Party of Canada. Dane Parker said that Roscoe had an exceptional memory and would quote extensively from varied sources during his speeches. Roscoe almost always dressed in work clothes and usually wore knee-high rubber boots. He spoke calmly and rationally, and people would listen. To emphasize an important point he would extend his right arm toward the crowd, with the stump of the forefinger he had lost as a child pointing in the air.

After their meeting on the Commons, the small band of radicals would often congregate at the home of a bootlegger, Pat Roddy, a poor but sympathetic World War I veteran who lived on the second floor of a shabby house on Market Street below the Citadel. The regular meetings of the Communist Party were often held at Roddy's on Sundays. Roddy sold a cheap but potent home-made brew called "seed beer" to the stevedores for five cents a glass and, with the proceeds, bought rum, which he sold to the socialists.

Dane Parker, the Halifax party secretary, decided to try to convert the dock workers who gathered at Roddy's to revolutionary socialism. He formed what became known as the Seed Beer Club, and every Sunday night he would lead discussions based on books such as *The Communist Manifesto* and Marx's *Value, Price and Profit*. Understandably, given the surroundings, Parker found it difficult just to keep the discussion on topic—winning recruits was next to impossible—and, disconcerted, finally disbanded the Seed Beer Club.

Another group that met at Roddy's was the Halifax branch of the Canadian Labor Defence League, established by the Communist Party to defend the many radicals being deported from Canada. Halifax was the main centre for government deportation of the thousands of foreigners it didn't want in the country. The deportees were often arrested in the night, quickly moved to Halifax, and given a quick hearing without a proper defence before being sent abroad. Roscoe bitterly complained about the manner in which the deportees were treated. In a letter published in *The Halifax Chronicle* he contrasted their fate with the "tender solicitude" shown by the RCMP toward a group of men charged with the crime of smuggling $5 million worth of liquor into the Maritimes and other parts of the country. The government "invited" the sixty-one people charged to attend a court hearing in Montreal and agreed to pay their travel expenses. When the case was heard in court, charges against the accused, who included Samuel Bronfman and his three brothers, were dismissed.

❧

The Depression gave rise to a new bitter enemy for Roscoe and other socialists. In early 1934 Roscoe wrote to *The Halifax Chronicle* warning readers that the Nazis in Germany had plans to extend their philosophy around the world. By that time Roscoe and other communists were among the first people in the Maritimes to speak out about the growing threat of fascism in Europe. They were anxiously listening to and watching out for the reports: Playing upon national pride and racism, fascism—with its roots in Mussolini's Italy—was putting power in the hands of an elite that ran the government and corporations and controlled society with the force of the militia. Fascism, they noted, was also turning out to be a negative reaction against the growth of Communism and democratic egalitarianism in several European countries.

Roscoe read *Mein Kampf*, Hitler's formula for restoring Germany and the

Aryan race to their former greatness and for regaining the lands lost at the end of the First World War. Hitler, it seemed to him, was some kind of a mad man, but one who bore watching. Concerned about what was happening, Roscoe wrote several times to Germany's acting consul-general in Montreal, protesting against the Nazi terror in Germany and demanding that the communists arrested by the Nazis be released or brought to trial in open court. Roscoe said it was likely they would have been killed had there not been public protest in many places in the world.

In Canada small fascist organizations had begun to spring up, including a group in Nova Scotia started around 1933 under the leadership of William Crane, an unemployed Halifax tailor. Restaurant owner Conrad Sauras was particularly concerned about the progress that Crane was making. Sauras suggested that someone follow Crane to attempt to find out where he was getting his support. According to Dane Parker, secretary of the Communist Party local, Crane was followed everywhere he went for two days. "A list of people was made; some doctors, some prominent people who apparently were sympathetic," Parker said. "The main people who were supporting Crane were all gentiles. They wanted to do away with the competition from the Jews."[14]

To combat the growing fascist sentiments in the Maritimes, Dane Parker, Rev. Grenfell Zwicker of the United Church, and other people set up a small, antifascist organization. It later became a branch of the League Against War and Fascism. To alert the public to the growing threat of fascism, Parker and Zwicker hosted a weekly radio program on Halifax station CHNS. They staged mock debates based on competing ideologies. However, the station management became concerned that the program's content was too left-wing. Often unable to say what they strongly believed—that the Nazis had the Reichstag burned, for instance—they would state some of their more controversial accusations in German, such as, "Goering brunt der Reichstag."[15]

❧

When the number of unemployed men roaming across the country grew well into the tens of thousands, the government set up what it called relief camps which were operated by the military—Roscoe called them slave camps. The Halifax Citadel Hill was turned into a relief camp where more than 350 men lived under strict rules and worked on road crews to earn their keep. Roscoe said that the men—many of them very young—were subject to military discipline of the Hitler labour camp variety.[16] One time when the men in the camp refused to work until it was agreed they would be paid twenty cents every day, no matter how poor the weather, twenty RCMP officers armed with riding crops entered the Citadel to evict any men who refused to go back to work. Another thirty city police, armed with batons, waited just outside the main gate. Seventeen of the men were evicted.

By 1934 and 1935, conditions for many families throughout the Maritimes were dismal. The three provinces were among the last to adopt such social programs as old age pensions and mothers' allowances. They were also on record as opposing unemployment insurance.[17] In New Glasgow a mass meeting called in the middle of winter to protest the refusal of town officials to set up relief programs was told that many people had no bed clothes and no underclothing. In Truro police reported a rash of house break-ins in which the main thing stolen was food. In Summerside, Prince Edward Island, three hundred men pleaded with town council to allocate money for work so they could buy clothing and books to send their children to school; they were told that council was prevented by technicalities from voting money for such purposes. These conditions existed even though huge amounts of food were being wasted. Speaking to the Halifax founding meeting of the Workers' Unity League in November 1934, Roscoe said wheat had been burnt and cattle slaughtered while the unemployed starved. He said that thousands of acres of potatoes were allowed to freeze in the ground in northern Maine because the price offered by the speculators was not enough to pay for the digging.

While all this was going on, the banks and many other businesses were still making huge profits, and a Royal Commission on Mass Buying was told about companies cheating customers by shortweighting their bulk purchases of groceries.[18] One of several horror stories told to the hearing was the case of Canadian Canners of Hamilton, which dominated the canning industry in the country. The company had reduced its payments to fruit and vegetable growers by 43 per cent between 1929 and 1933 while recording profits of more than $2.1 million.[19] In spite of its findings, Roscoe said that the Royal Commission's evidence of "slavery and thievery" on the part of large corporations was nothing new. "We have for years charged that these conditions existed, making life a hell for thousands of workers and their families, only to have those pious patriots deny the charges. Now they are on record." He predicted that the government would do little to stop the abuses "except to use the existence of the Commission as evidence that its heart bleeds over the wrongs of the workers. And both parties will fall back on the constitutional issue to prevent any action."[20]

The callous attitudes of the corporations and the do-nothing governments of the Depression renewed Roscoe's commitment to the Communist

Party. In February 1935 he got a letter from friends in New Brunswick asking him to come and speak to them—but further specifying that he could discuss the CCF or socialism, but not communism. They feared it would set their movement back. This stipulation set Roscoe off, and he wrote an angry response that he sent to the Cape Breton newspaper, *The Steelworker*: "They still toil to build up a house of cards based on hazy radical phrases that will collapse at the first crisis due to the self-seeking of their leaders and lack of fundamental knowledge of their membership. They are afraid of communism because to mention it might drive some of the mush-heads out."[21] He declined the invitation to speak.

❧

In the spring of 1935, party leader Tim Buck visited Nova Scotia following his release from Kingston Penitentiary. The incarceration of Buck and the other Communists had boosted the party's popularity across the country and would result in the party being made legal again in 1936. Buck was greeted by large crowds in Cape Breton. When the local council in New Glasgow refused to issue a permit to allow him to speak in one of the town's halls, he spoke to an overflow crowd of seven hundred people one morning at the Music Academy.[22]

When Buck spoke in Halifax at the School for the Blind, he was greeted by a standing-room-only audience that included a provincial cabinet minister and several of the city elite. Speaking to a hushed audience, Buck described some of his most dramatic experiences—including what he was convinced was an attempt to kill him during a prison riot—during the nearly three years that he and seven other Communist leaders had spent in the penitentiary. Following the event, Roscoe—the only other speaker that night—and the others took Buck to Pat Roddy's, and a friendship began to develop between Roscoe and Buck. Roscoe, like many others, found Buck to be a charming, likeable, and courageous person.

Soon after Buck's visit to Halifax, it seems, Roscoe became a member of the central committee of the party.[23] Still, friends considered Roscoe to be independent of the party in many ways. If Roscoe felt it necessary, he would blast party officials in private for things he thought they were doing wrong—although he wouldn't speak out publicly against the party. Some members of the party questioned Roscoe's membership because he was a small businessman and the party opposed private ownership and the employer-employee concept. That matter was dropped when it was decided that Roscoe's general contribution to communism far outweighed the fact that he ran a small, struggling business.

❧

In 1934 Roscoe added to his political endeavours by starting to write a newspaper column for *The Steelworker*, the Cape Breton weekly. The flamboyant paper, with a circulation of only a few thousand copies, supported the Communist Party position on most issues. Even though only a small percentage of the circulation was outside of Cape Breton, the column gave Roscoe his only outlet for writing beyond his letters to newspapers, which were often either heavily censored or rejected because of his strong views. *The Steelworker* was published by M.A. MacKenzie of Sydney, a colourful, outspoken man who had been blacklisted by the steel company for his involvement in unionizing efforts.[24] The fact that Cape Breton supported two radical weekly papers during much of the 1930s was a sign that the island was by far the most radical area in the Maritimes, especially given its militant workers in the coal mines and the steel mill of Sydney. The other left-wing paper, *The Nova Scotia Miner*, was edited by J.B. McLachlan until it closed in 1936.

The Steelworker, which became the *Steelworker and Miner* in 1939, was a small-budget operation. When the paper was short of money, MacKenzie was not above letting readers know that he had a family of ten children to support. The masthead slogan read: "An Island of Truth in an Ocean of Lies." Roscoe's columns appeared in *The Steelworker* from 1934 until 1936 and again from 1943 until 1953, when the paper ceased publication. Over the years he wrote more than 250 columns. Other contributors included Roscoe's friends Joe Wallace and John Mortimer. Mortimer came from the Miramachi area of New Brunswick and had been a missionary in India. Retired and living in Sackville, he wrote primarily on international affairs for *The Steelworker*, under the pseudonym "Kentucky Colonel."

Roscoe and the other *Steelworker* writers were not, of course, paid for their work. Like his other radical activities, Roscoe's column was a labour of love. Roscoe had no pretensions about being a serious writer or an intellectual—he detested intellectuals. His writing usually reflected his outright anger at the world's inhumanity, and he hammered away at various evils, from fascism to the Canadian establishment or, later, what he believed to be the American quest for world domination. He tended to be blunt, writing about issues in a black-and-white manner, and a dogmatic faith in radical socialism underlined most of his writing.

Roscoe's columns dealt mostly with international political issues, but he occasionally tackled other topics. In one of his first columns, written in the fall of 1934, he wrote about protests over a new high-school textbook called *The Story of Civilization*, which explained the theory of evolution. In Sydney, some twelve hundred adherents of the Presbyterian Church had signed a petition condemning the book. Dr. Alexander Murray of Sydney said the book "contained a dogmatic declaration that man had come from 'brute ancestry.'"[25] Fundamentalist ministers attempted to discredit the work of Charles Darwin, and their comments appeared on the front pages of the province's newspapers.

Roscoe responded by pointing out the similarities between the fundamentalist backlash in Nova Scotia and the Scopes "monkey trial" in Tennessee in 1925. He described the hypocrisy of the southern fundamentalists, who preached an inflexible morality and yet condoned the racism and violence of lynch mobs. He contemptuously dismissed the close-mindedness of those who felt that knowledge came from revelation instead of research and experimentation. Roscoe urged people to refuse to sign petitions being circulated by the clergy. He defended *The Story of Civilization* not only because he was offended by those who opposed it, but also because he felt it was a good textbook.

> In the past our children have been taught history that was a pack of lies and jingo ravings. The doings of lecherous and cut-throat kings and lords and ladies who rode on the backs of our forefathers and enslaved them were dished up as history. According to these accounts, the progress of the human family was due to the accidents, mistakes, wisdom, virtues and miracles of these beings. History as taught reflected very little of man's discoveries and achievements, and never so much as hinted that progress was the result of the struggles carried on by various classes at various times, against the tyranny of kings, landlords, etc.[26]

Roscoe could find a political angle to just about any story. Around this time the Dionne Quintuplets were born to a poor French-Canadian family in Callander, Ontario—the first quints in the world to survive infancy. Astounded at the care and attention lavished on the five baby girls, Roscoe suggested in his *Steelworker* column that all the children of Canada should be given the same care. He wrote that while the politicians of Ontario were making so much fuss over the quints—and reaping publicity for the province at the same time—there were thousands of other children across the province who were undernourished, sick, and neglected.

❧

Partly due to his column in *The Steelworker*, Roscoe was asked to go to Cape Breton in the fall of 1935 to support J.B. McLachlan's campaign for election to the federal Parliament. Roscoe was pleased by the invitation. He had not often been able to go to Cape Breton since the Fillmores had settled in Centreville. It was nearly 350 kilometres from the Annapolis Valley to Glace Bay, the centre of radicalism on the island, and there was infrequent communication between the island communists and the small Halifax and Valley groups that Roscoe associated with.

While the island had never elected a communist to a federal or provincial seat, it had elected several reformers who ran as labour candidates. A Farmer-Labour group including four Cape Bretoners had provided the main opposition in the Nova Scotia legislature from 1920 until 1925.[27] McLachlan had run for office under various labels four times, without success.

Roscoe and McLachlan were not close friends, largely because of the distance that separated them. But Roscoe had admired McLachlan's contribution to the cause of revolutionary socialism ever since the two men first met as members of the Socialist Party in 1910. McLachlan, sixty-six at the time of the election, was editor of *The Nova Scotia Miner* and had been instrumental in setting up the Amalgamated Mine Workers of Nova Scotia (AMW), a predominately communist union that had broken away from the UMW. In 1933 he had been named president of the Workers' Unity League (WUL), an organization that aided the unemployed and set up communist unions as an alternative to traditional pro-capitalist unions.

The people of Cape Breton had been hit hard by the Depression. The economy of the island was dominated by a consortium that owned the coal mines, the steel mill, and most of the shipping. Jobs had been cut and many of those who were employed at all worked only two shifts a week and earned barely enough money to stay alive. The state of poverty led to some ugly and violent incidents during the months preceding the 1935 federal election. In North Sydney three hundred people who rioted after being told their relief was being cut by 25 per cent were quelled by police using whips. In June nearly one hundred families in the Glace Bay area were destitute and without food after the government said it had no money for relief payments.[28] The Communists maintained that conditions would not improve while the traditional capitalist political parties were in power. As an illegal organization until 1936, when Section 98 of the Criminal Code was abolished, the Communists risked further persecution by running thirteen candidates in the 1935 election.

The Liberals were favoured to take the riding McLachlan had selected, Cape Breton South, but if anyone were capable of an upset it was J.B. McLachlan, who was something of a father figure to many of the workers on the island. Tall and gaunt, easily recognized by his walrus moustache and balding head, McLachlan was a capable and knowledgeable speaker.

Addressing his audience in a guttural Scots brogue, he drew large crowds during the campaign.

More than 250 people heard Roscoe speak in support of McLachlan at the Strand Theatre in Sydney in early October. He told Cape Bretoners that because of what he had seen in Canada since the Depression began, he was convinced it was useless to appeal to those in power to improve conditions for poor working people. He urged workers to vote for McLachlan so the new member could carry the class struggle to the House of Commons.[29]

With support from the likes of Fillmore, Tim Buck, Annie Buller, Sam Scarlett, Jack MacDonald, Arthur Evans, and Joe Wallace, McLachlan's campaign did well, surprising many people, including the officers of the RCMP. "The movement is gaining a lot of strength," said one RCMP report. "McLachlan is keeping clear of knocking the church and catering for a respectable vote."[30] The Cape Breton establishment must have been worried: The day before the election a prominent parish priest condemned communism in a sermon reported on the front page of the *Sydney Post-Record*. He said no Catholic worthy of the name could conscientiously support the standard-bearer of communism, the deadly enemy of religion.

McLachlan didn't win, but he polled more than 5,300 votes and came third to the winning Liberal candidate. He had majorities in most working-class wards in Glace Bay, and received by far the largest vote ever recorded by a revolutionary socialist in the Maritime provinces. The election was McLachlan's last. Increasingly embittered by the Communist Party's campaign to force his Amalgamated Mine Workers of Nova Scotia to join the mainstream United Mine Workers, he resigned from the party in 1936. Following his lead, other Cape Breton radicals also left the Communist Party. In the 1937 election many Communists, in keeping with the party's "united front" policy, worked for the new Cape Breton Labour Party, which soon became a part of the CCF. McLachlan died of tuberculosis in 1937. While the Communist Party continued to operate in Cape Breton, it never again played as prominent a role in the politics of the island's labour movement as it had during his era.

FOURTEEN

Fish Plants and Workers' Rights

A PARTIAL RECOVERY of the economy in many parts of Canada in 1936 restored business profits to almost pre-Depression levels.[1] Still, the men who owned the factories, mines, and mills and who controlled the economy were no more benevolent toward workers than they had been during the worst of the Depression, and working-class incomes were still below their previous levels.

In the Maritimes most workers lacked the clout to force their employers to make wage increases. Cape Breton miners, who had seen their wages cut by 12.5 per cent in 1932, had to stage slowdowns and strikes until Dominion Steel and Coal—which had a profit of $1.3-million in 1937—restored half the earlier wage cut. In Minto, New Brunswick, the United Mine Workers led a fight to gain union recognition and higher wages for coal miners, but during the winter of 1937-38 the coal operators starved the workers into submission while the provincial government watched from the sidelines.

Many jobs in the Maritimes were strictly controlled for political reasons or for gain by unscrupulous individuals. George MacEachern, a prominent Communist who worked at the steel mill in Sydney, said that jobs at the steel mill were sold for profit. Often when someone arrived in the city and wanted work at the mill, the newcomer would be directed to a handful of people employed by the company who could find a job, for a certain price. Each ethnic group in the city had middlemen who could secure a job for payment. Men desperate for work, with families to feed, would be asked to pay about $75 for a difficult part-time job shovelling coal at the mill's coal bank and even more for a better full-time job inside the mill. When the men couldn't pay, they often didn't get work.[2]

In the Annapolis Valley many jobs were controlled by the ruling provincial government through a patronage-ridden system. One of the biggest plums was the Nova Scotia Sanatorium, located in Kentville. The staff was subject to wholesale firings and hirings at every change of government. Contracts for services at the "San" were also politically controlled. As in other parts of the province, jobs on highway were also strictly controlled by the local political hierarchy through Standard Paving. When George Nowlan represented the Valley riding in the provincial legislature during the Depression, he proudly commented that the local paving-company boss phoned him every day to ask about men who should be given work. On election day, many votes were bought by the promise of a job or a favour.

In Halifax in the late 1930s the patronage system was controlled by the ruling Liberal Party. Roscoe Fillmore's Communist friend Dane Parker, who was a member of the Halifax District Trades and Labor Council, recalled that to get a job people often had to pay a key Liberal who held a senior position in the union that controlled the job. In many cases the worker then had to continue to pay to keep the job. While the heads of the unions would change when the Tories were elected, the patronage continued unabated. Such practices applied to some of the unionized workers in Halifax, particularly the longshoremen. The majority of workers living in the city were unorganized and paid pitiful wages.

The small Halifax Communist group began to talk about launching a drive to organize some of the poorly paid, exploited workers in the province who were being ignored by the traditional labour movement. They knew that elsewhere in the country Communists were making a major contribution to the growth of unionization. In Cape Breton, Communist George MacEachern was playing a major role in the organizing of the Sydney steelworkers. Parker and the others felt it was time to continue this trend in their city. They also knew that if they were ever to exert any political influence they would have to win the respect of the working class. Parker argued that the best way to win this respect was to help the workers form strong unions so they could win better wages and working conditions. The Communists hoped to offer an alternative to the patronage that controlled so many jobs in Halifax and elsewhere in the province. When Roscoe was in Halifax to sell nursery stock at the market he would meet with Parker and the handful of other Reds and they would discuss ways of launching an organizing drive. While Roscoe was extremely supportive of the new Communist initiative, he was unable to play much of an active role because of his responsibilities at the nursery.

The Halifax group recognized that women workers were among those most neglected by the labour movement as well as the poorest paid and most powerless members of the labour force. It is intriguing that the first two unionization efforts of the Communists were on behalf of women workers, who were notoriously difficult to organize. As it turned out, both of these efforts were failures. Women workers at Moir's Chocolates in Halifax rejected Parker's efforts to establish a militant organization because they were afraid of losing their jobs. Parker then switched his attention to nurses working at the Victoria General Hospital in Halifax. Nurses were poorly paid, they worked long, difficult shifts, and the ones in residence lived in deplorable conditions. After several secret meetings and evenings spent with the families of some of the nurses, Parker concluded that they couldn't be organized because, like the Moir's workers, they were terrified of being fired. For low wage earners the consequences of being fired were more serious than for other workers.

Parker turned his attention to the fishing industry. Fishing was the largest employer in the Maritimes, with about forty thousand people involved in different aspects of the business. There was both offshore fishing, where large trawlers would stay at sea for several days at a time, and inshore fishing, where smaller boats would fish closer to shore, returning to port every evening. Throughout the region there were fish-processing plants where workers—many of them women and teenagers—were employed cutting and packaging various fish products.

Conditions in the small fishing villages were deplorable. Roscoe wrote in *The Steelworker* that fishermen were risking their lives daily for an amount of pay that barely allowed them to fill their gas tanks and go looking for more fish. He said that women and young girls were working in fish factories for seven and a half cents an hour, ten hours a day, and then they had to pay their own room and board. The children of such families, Roscoe wrote, were ill-nourished, poorly clad, and poorly educated. They were the slaves of the wealthy fish companies.[3]

The 1935 Royal Commission on Price Spreads, which investigated charges of excessive profits by companies during the Depression, found that many of the fishermen were so poor they couldn't afford to equip their boats or buy proper fishing clothing for the 1934 season. Roscoe wrote that Lunenburg fishermen were paid less than one cent a pound for fish that were sold for up to twenty cents a pound in Kentville.[4] The Commission said the low incomes were largely the result of the fish companies using centralized buying and price fixing to destroy the bargaining power of fishermen.[5] The few companies that processed and marketed most of the fish were making handsome profits.

There were also other ways to make a profit from the hard work of the fishermen. Ralph Bell, a Halifax businessman and speculator, bought up several fish companies during the late 1920s and early 1930s. After establishing a monopoly in some communities, Bell sold the plants in 1936 at a huge profit. Roscoe had a special dislike for Bell. He accused Bell of supporting fascist beliefs, arguing that the financier had distributed thousands of reprints of a convocation address by Dr. George Barton Cutten, former president of Acadia University, that was "pure fascism." According to Roscoe, the speech condemned all social legislation, declared society was weakening itself by helping the weak to survive, and hailed the "successful man" as the flower of man's struggle all through the ages. "Mr. Bell and his kind are 'flowers'—flowers of capitalism, of poverty and disease, malnutrition and ignorance," Roscoe wrote some ten years later in the *Steelworker and Miner*.[6]

During the 1930s, co-operatives were being touted as a way to ameliorate the poverty of the fishing villages, but Roscoe argued that these were useless, childish schemes, controlled more or less by the clergy. He said that politicians and newspaper editors praised the puny efforts of the co-ops, which saved the fishermen pennies, while the big thieves, the speculators and bankers like Bell, made huge profits running the much larger fish plants.

❧

In 1937 Dane Parker decided that working conditions in the two fish plants located on the Halifax waterfront were so bad that the employees might be prepared to fight for a union. About 375 people worked as fishhandlers and cutters for two companies, General Seafoods and National Fish. General Seafoods was unaware that its newly hired production manager, twenty-one-year-old Charlie Murray, was a Communist. The son of a Presbyterian minister, Murray had been strongly influenced by the poverty of the Depression and had just attended a four-month Communist Party school in Toronto. In his new job Murray did what he could to help the unemployed workers who drifted through Halifax. Working as a biologist, he would occasionally mutilate fish so he could take them home. His apartment was left unlocked and he let it be known that it was a place where people who were hungry could go and prepare for themselves a meal of salt fish and potatoes.[7]

Murray helped Dane Parker get a job in the fish-freezing section of General Seafoods so Parker would be well positioned to organize the workers. Murray also used his position inside the company to feed information about the firm's strategy to Parker. The other key man in the organizing effort was Alex "Scotty" Munro, a rugged fishworker who had migrated to

Canada from Scotland as a fourteen-year-old orphan to work on a farm. To survive during the Depression, Scotty had become a prizefighter. Munro's job at one of the plants was the exhausting and dirty work of throwing fish out from the hold of a ship with a pitchfork.[8]

For Parker and Munro it seemed a perfect time to gain recognition for their new Fish Handlers Union. The provincial legislature was passing a bill that would, for the first time, permit the workers of Nova Scotia to join the union of their choice and bargain collectively. The legislation was introduced as the result of lobbying by the Sydney steelworkers, who could not get their company to voluntarily recognize a union. As a result of the new legislation, sixteen different unions representing more than four thousand workers were starting to go through the process of gaining recognition. Premier Angus Macdonald hailed the legislation as a model for other jurisdictions in North America.[9]

In August 1937 the Fish Handlers Union and one of the companies, National Fish, began to discuss working conditions. When the company refused to come to terms with the union, the workers refused to unload fish from the trawlers that docked regularly at the fish plant. Saying there was no work, the company responded by firing sixty workers, including most of the employees active in the union drive. The Fish Handlers Union escalated the dispute by taking its members out of the National Fish plant and putting up the first picket line that Halifax had seen in twenty-five years. However, some non-union people continued to unload and process fish from a trawler. *The Halifax Chronicle* quoted Dane Parker as saying that the union should have tried to stop the first scabs from entering the plant: "If we had given one or two of them a good mauling, no others would have showed up."[10]

After a week on the picket line the workers began to lose patience. They had been trying for more than eight months to get recognition under legislation that had been hailed as a boon for working people. On Sunday, January 16, 1938, the workers, hurling rocks and armed with clubs and barrel staves, attacked National Fish buildings. No serious damage was done, but the action led to an emergency meeting of Premier Macdonald and union and company officials. An agreement was reached to allow the National Fish workers to vote on whether they wanted a union.

All workers, including many who had previously been fired, returned to work to await the vote. Nine days before the vote, a headline in the newspaper proclaimed that the two fish companies were considering moving to Newfoundland, where they could enjoy better tariff regulations and more agreeable labour conditions.[11] Had the workers voted months earlier, and had strikebreakers not been allowed to vote, the union almost certainly would have won. But after the intimidation by the company, the vote was narrowly lost, 135 to 125. The strike was defeated, and the province's Trade Union Act had failed the Halifax fish-plant workers.

❧

The fact that Parker, Munro, and Murray were Communists was not raised publicly as an issue in the Halifax strike, but with the coming of the Second World War and Russia's non-aggression pact with Hitler communism became an issue in another drive to organize fisheries workers on the south shore of Nova Scotia.

The 1939 campaign to organize two fish plants at Lockeport, a village of fourteen hundred people, occurred not as the result of an organizing effort initiated by the Halifax Reds, but because of a dispute between the Communist-led Canadian Seamen's Union (CSU) and Canadian shipowners. The owners wanted to prevent the CSU from organizing their ships on the Great Lakes, and one of their plans was to hire Nova Scotia fishermen to work as strikebreakers. When the CSU learned of the owners' plan it launched an organizing campaign among Maritime fishermen. Pat Sullivan, president of the Seamen's Union, and Charlie Murray, who had been recruited by the CSU to help organize the workers, travelled to many small fishing communities along Nova Scotia's south shore and found the men eager to unionize.

The two fish plants where much of the south-shore catch was processed were located in Lockeport. Both companies, Lockeport Company and Swim Bros. — the two main employers in Shelburne County — were strongly opposed to the idea of a union. While many fish-plant workers and fishermen were living in poverty, the fish companies, as Sullivan argued, were enjoying huge tax exemptions from the provincial government.

In August the Canadian Fishermen's Union (CFU) was organized with two locals, one for fishhandlers and the other for fishermen, with about 650 members. An RCMP report noted that Roscoe was among a group of men who had promoted the "trouble" in Lockeport. RCMP intelligence notwithstanding, Roscoe was not involved in the day-to-day workings of the dispute, although he frequently gave advice to Sullivan and Murray.

The union filed for recognition under the Trade Union Act but the companies refused to talk to the organizers. When the men showed up for work one morning in October they found signs posted on both fish plants indicating they were closed until further notice due to union activity. In an attempt to side-step the labour legislation, the workers were all fired.

The Nova Scotia government co-operated with many unions that sought recognition under the new Trade Union Act, but Premier Macdonald would have nothing to do with anything suspected of being communistic. Attempting to get the fishermen to rebel against the union, the premier and labour minister Lauchie Currie issued a statement claiming that Sullivan and Murray were Communists and that their main interest in forming a union was to promote the causes of the Communist Party. Sullivan and Murray, perhaps fearing that if they told the truth they would lose the support of some of the workers, stood before a union meeting in Lockeport and denied they were Communists.

Premier Macdonald was indebted to a small group of wealthy Nova Scotia businessmen headed by Ralph Bell for his start in politics. In 1932 Bell, who had invested heavily in the fishing industry, convinced a dozen other men to put up $1,000 each to back Macdonald's entry into politics. A year later, after Macdonald became premier of Nova Scotia, Bell boasted to friends, "We put Angus in the House." Bell and other businessmen wanted Macdonald to make sure there was as little interference as possible with their business interests.[12]

The real issue in the Lockeport dispute was not communism, but whether the workers were going to be able for the first time to have a tough union. What followed during the fall and winter is one of the most remarkable stories in the history of the Maritime labour movement.[13] When the company refused to recognize the union, the fishermen, who had been locked out and desperately needed to earn a living, took out an option on a closed fish plant and opened their own fish-processing co-operative. Experts from the Cape Breton co-op movement came to Lockeport to teach the fishermen how to run the company.[14] The co-op soon had twenty small boats delivering fish to it and was handling twenty-five thousand pounds of fish on good days. It sold the fish to miners and steelworkers in Cape Breton and to a company in Montreal that agreed to buy fish in large quantities.

At the same time the union put up picket lines on the road and the railway track leading to the fish companies to try to stop them from handling any fish. Most of the trains were stopped, but some trucks carrying fish broke through the lines. In late November the provincial attorney-general declared the pickets had no right to try to stop the company from marketing its fish. Within a few days the companies announced they would attempt to fully operate their plants, and the attorney-general sent sixty Mounties to the town to maintain law and order. A picket line at the plants was joined by fishermen from many villages along the coast who now belonged to other locals of the new union.

On the morning of December 11, seven hundred men, women, and children picketed the gates of the plants and the train tracks that led into them. They were expecting a train that was coming to pick up a load of fish, and as it approached they realized it was being led toward the town by fifty Mounties marching in two columns. The Mounties ordered the crowd to disperse. When the pickets refused, the Mounties pressed forward. The pickets fell back a few feet, wavered for a moment, then stood firm. The people of Lockeport knew that if the fish were to be removed from the plants the workers' struggle would be over. They would be forced to give up the union, and many of them would never again be able to secure employment in their home town.

Standing rigid and tense, the inspector commanding the RCMP shouted at the pickets to disperse in the name of the King. "It was then that the 'battle' got under way in earnest," *The Halifax Chronicle* reported. "The Mounties swarmed against the pickets and attempted to push the men and women from the tracks." The RCMP began to forcibly remove the women from the tracks. Several persons were slightly injured. "With the police vastly outnumbered by the picketers, one of whom climbed a box car and hoisted the Union Jack while others sang 'God Save the King,' 'O Canada' and other patriotic songs, Conductor 'Happy' Lennox communicated with railway headquarters at Bridgewater and the train was ordered to proceed from the town."[15] The RCMP marched away from the scene. The picket lines had held.

Premier Macdonald dispatched a hundred additional RCMP officers to the area and ordered the union and the companies to attend a meeting in Halifax. Despite the determination of the workers, the meeting resulted in a proposal that fell far short of the union's goals. The companies said they would recognize the fishhandlers' union but would not negotiate with "outsiders"—meaning Sullivan and Murray—who were not employed by the company. The fishermen—legally defined as "co-adventurers" rather than workers because they owned their own boats—could not be employees and therefore were not officially recognized as a union. Reluctantly, the men voted to accept the agreement on the recommendation of the union leadership. But two months after the settlement the workers declared that the companies had still not recognized their union. Only half of the former employees had been rehired, and the union complained of constant harassment. A more positive result of the new militancy in Lockeport was that in the town's next election, a new pro-labour mayor and council were voted into office.

The agreement meant that the union was free to bring fishhandlers elsewhere in the province into the union. Many locals had already been formed. The organization, which changed its name to the Canadian Fishermen's and Fish Handlers' Union (CFFHU), made plans for further union activity throughout Nova Scotia. But Charlie Murray and union president Ben MacKenzie received letters from labor minister Currie warning them not to be militant in their demands or they would face the consequences.[16] War had broken out during the lockout, and it wasn't long before Murray and Pat Sullivan were arrested and interned along with several other prominent Communists.

❧

In 1937 the nursery business in Centreville was still no more successful than it had been at the beginning of the Depression. But the fact that three of Roscoe and Margaret's children now had left home eased some of the financial pressure. Their oldest son, Dick, had returned to Nova Scotia from Massachusetts in the early 1930s and was living nearby and working at one job or another. When the eldest daughter Ruth had found little to do but help around the nursery, it seemed best for her to go live in Detroit, where she would be able to find work. So, in 1933 at age twenty, she had reluctantly left Centreville to go to live with Mabel and her family.

Once Ruth had settled in Detroit, that city became the prime destination for Roscoe and Margaret whenever they could spare the money to make the trip. After Ruth gave birth to two children, David and Barbara, the city became the Fillmores' home away from home. Their annual trips to Detroit, and Roscoe's many letters to his new grandchildren, were aspects of Roscoe's growing interest in maintaining a closer relationship with his family.

At age sixteen Rosa had married Gerald Skinner, who was running his family's grocery store and post office in a nearby village. Gerald and Dick Fillmore ran a wholesale fruit business for about two years until it went bankrupt. In 1935 Rosa gave birth to her first child, but the baby suffered from a heart ailment and was born with a rib cage and feet that were not fully developed. The little girl, named Lorraine, required special medical attention and spent several months in the Children's Hospital in Halifax. As Lorraine grew older the family found that she would not be able to walk. Eventually the doctors determined that if she were ever to have a normal life she would have to undergo an expensive operation in New York to fix her feet. The operation would cost almost $2,500, a sum that was far more than what Roscoe earned in a year. Everyone in the family, including relatives in New Brunswick and the United States, contributed until the amount was finally reached.

When Lorraine was four, Rosa and Roscoe took her to New York for the operation. The surgery was a success, but it would be a long time before the little girl would be well enough to lead an active life. During the period when she was back home and undergoing further treatment, Roscoe lavished attention on her. He would put her in a beer carton on the front seat of his truck and take her with him everywhere. Together they would drive the roads of the Valley, laughing and talking as Lorraine built her confidence for the difficult life that lay ahead. Later, after a long recovery, she attended school, graduated, and became a secretary.

The loving care that Roscoe showed for Lorraine brought Rosa and Roscoe closer together. Rosa felt that she finally got to know her father well—there was a compassionate side to his nature she had not seen before. Previously his fiery temper had made her afraid of him. To Rosa it had seemed that Roscoe had an air of male bravado that covered up his real feelings of love for all his children.

With the new opportunity for long talks with her father, Rosa also detected what she believed to be a major change in Roscoe's approach toward politics. When she was living in Halifax in 1938 so Lorraine could get special treatment at the Children's Hospital, Roscoe and Tim Buck spoke to a small group of Communist friends during a private gathering at Conrad Sauras's restaurant. Roscoe spoke calmly and clearly, with humour, about the problems people faced during the Depression. It seemed he had begun to realize he couldn't browbeat people into accepting his view of the world. It was a lesson that Jake Resnick, the Cumberland County merchant, had tried to teach Roscoe much earlier when the two men became friends during the time of Springhill's long strike.

The two children still living at home in 1937, sixteen-year-old Alex—or Allie, as she had come to be known—and Frank, thirteen, had no objections to Roscoe's politics. In high school Allie expressed socialist opinions, and later, after the outbreak of the Second World War, she attended antifascist meetings in Kentville. After graduating from high school she took a job as a sales clerk in Stedman's Department Store in Kentville.

Frank's main interests while growing up seemed to be playing hockey and, according to schoolmates, getting into trouble. Like Ruth he had bright red hair and a quick temper. Even though Roscoe would harangue Frank for his misdeeds, he would always pay to fix anything the boy might break or damage. Even Frank's introduction to radical politics got him into trouble. He and his pal "Haddie" Porter were sitting at the front of a packed hall in Centreville when speaker Annie Buller held up a large picture of what she said was a group of Russians. She was explaining to the audience how happy they looked. "Look," she said, "they're all smiling." Popping up from his chair, Frank announced that he had seen a picture in a magazine just a few days earlier of a bunch of Germans, and they were all smiling too! Someone in the audience yelled, "What the hell are those kids doing in here?" and Frank and Haddie were grabbed by their collars and hustled out the door.

Despite his many escapades, Frank was the one of all of Roscoe's children who showed the most interest in radical politics. In school he would often defend Roscoe's left-wing views, and he would become the only one of Roscoe's five children to join a communist party, the Labor-Progressive Party.

❧

Toward the end of the Depression an unexpected development involving Dick resulted in Roscoe getting a well-paying job away from the nursery. Dick, who had considerable nursery experience, was head gardener with the Dominion Atlantic Railways (DAR), which operated the railway line through the Annapolis Valley between Yarmouth and Halifax. The job included caring for one of the railway's prized possessions, Grand Pre Memorial Park, which had been opened on marshland near Wolfville during the First World War in tribute to the thousands of Acadians deported from the Maritimes during the expulsion of 1756.[17]

Dick enjoyed his work with the DAR, but he gave up the job in 1938 to attend Acadia University. The university had been unable to attract enough students during the Depression, and the president was trying to boost enrolment. He heard about Dick and drove down to Grand Pre Park to offer him a chance to attend Acadia in return for the upkeep of the lawns and gardens at the university. Dick eagerly accepted the offer, even though he had to attend high school for two years and receive special tutoring before he was ready for university.[18] His decision to go to Acadia opened the door for Roscoe to take over his son's position as DAR head gardener. The work included the propagation, planting, and care of lawns and gardens at eight sites, including Grand Pre. In winter Roscoe grew nursery stock in a greenhouse located beside the Kentville train station.

By far the most important of the parks was Grand Pre, where the Acadians had farmed more than 250 years earlier. The grounds at Grand Pre were steeped in Acadian history. Near the church stood a row of willows planted by the Acadians. Behind the church were apple trees that the settlers had brought from France. Many of the flowers, including rose bushes, had been sent from France by descendants of the Acadians. Perhaps the park's most striking feature was the ivy-covered replica of the Acadians' original Grand Pre church, built in Norman-style architecture. Inside the church were many artifacts from Acadian times, including spinning wheels, furniture, and old coins. Located directly in front of the church, a life-size marble statue of Evangeline watched over the meadows and homeland of her ancestors. A reconstructed Norman gatehouse stood at the entrance to the park, and off to one side was the railway line.

Under Roscoe's direction Grand Pre Memorial Park became one of the most popular tourist sites in the Maritimes. The poet Longfellow had immortalized the hardships faced by the Acadians in his poem "Evangeline," and each year thousands of tourists—including many Acadians from Louisiana—visited the park to relive a dramatic part of Maritime history. Every spring there was a flurry of activity to get the park ready for the closing ceremonies of the Apple Blossom Festival, one of the Valley's most prestigious social events. The queen of the festival was always crowned in front of the statue of Evangeline.

Large bright-coloured beds of marigolds, petunias, and other annuals were planted to attract tourists. Shade trees were perfectly trimmed, and the lily ponds, where a flock of ducks spent the summer, were carefully groomed. Above everything else, Roscoe was concerned about the care of the lawns, and he made sure they were groomed to perfection.

The same summer that Roscoe got the job with the DAR, the Fillmore home in Centreville was extensively damaged by fire. Early on a Sunday morning in June 1938, a neighbour saw flames shooting from the Fillmore home and rushed to the house, calling out to members of the family. Roscoe and Frank were in Halifax but Margaret was at home. She escaped, and as the fire swept through the house she and a neighbour calmly removed armload after armload of Roscoe's books from the house. Someone called Roscoe's close friend Jimmy Sim, who lived three kilometres away, and Sim soon arrived with a team of horses pulling a spray rig. The Kentville Fire Department, which had been out on another call, also arrived, but by this time the house had been gutted.

Roscoe and Frank arrived home later that day to find most of the walls had been burned out and there was a gaping hole in the roof. They never did find out what caused the fire, though it seemed to have started in a shed next to the house. The loss, estimated at $2,500, was not covered by insurance.[19] The family draped a huge tarpaulin over the hole in the roof and were able to continue living in the house.

❧

Late in the summer of 1938 Tim Buck visited the Fillmores in Centreville. He gave an outdoor talk in Centreville that, despite cool weather, held an audience of more than two hundred people until after 11 p.m. His speech was warmly applauded. In his talk, Buck urged the formation of a united front of political parties to replace the old-line parties and said that the government should provide aid for the fishing and mining industries, which he said were being exploited by large business interests. It was the first time that a major figure in the Communist Party had ever spoken in the Annapolis Valley. *The Advertiser* reported that Buck, slim and smiling, completely captured his audience with his fluent oratory and masterly diction.[20]

By that time many Canadians were hearing reports that not all was well with the Communist Party and its leadership in Moscow. There were stories that Joseph Stalin had become a dictator, that people who opposed Stalin were being purged, and that many peasants and farmers who didn't like the Communist government had been imprisoned or killed. While many of his neighbours undoubtedly believed the horror stories coming out of the Soviet Union, Roscoe felt that all criticism of the Soviets was Western propaganda, just like the propaganda he had seen right after the revolution. During the fall of 1934, for instance, many critics of the Soviet Union were claiming that the country was again being plagued by famine. But when *New York Times* reporter Harold Denny visited the alleged famine districts and reported he could find no famine, Roscoe gleefully reported Denny's findings in his column in the *Steelworker and Miner*.

Proud of the progress that was being made in the Soviet Union, Roscoe frequently cited articles from mainstream newspapers such as *The New York Times* that praised the achievements of the Soviet Union in such areas as health care and education. "We know that Russia is no Utopia," he had written in *The Steelworker* in 1935, "but this we can boast, that in that country—as large as the whole North American continent and with a great population—there is not a hungry, malnourished or badly clad child."[21]

While Roscoe was correct in claiming that there had been great achievements in the Soviet Union, he was on less solid ground on the few occasions when he talked about oppression in the country. He wrote in *The Steelworker* in 1934 that those people who talked about the Stalin dictatorship had "swallowed a bunch of hokum."[22] At the time hundreds of thousands of people were being forced to give up the method of agriculture that they had practised for generations to take part in collective agricultural projects. Reports indicated that people who resisted these changes were subject to reprisals. In 1933 *The New York Times* and other mainstream publications had warned that Stalin's second five-year plan, which would send more than two million people from Soviet cities to participate in agricultural reform and industrial development, would cost many people hardship, suffering, or even their lives—criticisms that Roscoe did not believe. Only much later, in his personal autobiography written in 1959-60, would Roscoe acknowledge that the Soviet Union, due to fear of a persistently hostile world, had "undoubtedly dealt harshly with the slow moving peasantry."

Dane Parker later recalled that none of his communist friends would believe the stories about all the killings because of all the other lies circulating about the Soviet Union. He said that during the late 1930s one of the standard jokes among them was about how Roscoe was keeping a record of all the people that *Reader's Digest* claimed had been killed in Russia. "It added up to hundreds of thousands," Parker said. "It was impossible!"[23] Sadly, though, many thousands were dying.

Perhaps the only thing that might have changed Roscoe's mind about some of the events in the Soviet Union would have been if someone of the stature of Tim Buck had said there were problems. Buck was one of a few Canadian Communists who had first-hand knowledge of some of the developments taking place in Moscow. The year before his visit to Centreville he had been in Moscow for party meetings and had heard that his friend Nikolai Bukharin—a man whom he greatly respected—was going on trial for treason, charged with taking part in a conspiracy against Soviet authorities. Buck attended the trial and it quickly became clear to him that while Bukharin was guilty of joining a rival faction, the fact that he thought this necessary indicated that there were serious problems in the USSR. Buck was in a turmoil. Over and over in his mind went the question: If Nikolai Bukharin joined an opposition group there must be something wrong.[24] Following Buck's return to Canada, there was discussion of the way that opponents of the regime were being dealt with, but because the Canadian Communists were more concerned about the growth of fascism in Europe than what was going on in Moscow, nothing was done.

If Buck discussed his personal concerns about the Stalinist regime with Roscoe and Jimmy Sim when they met privately, none of them gave a thought to speaking out publicly about their fears. Despite what he was told, however, incidents occurred in Halifax that must have led Roscoe and others to ask questions about the nature of official Communism. The Communist Party in Canada had itself become authoritarian and inflexible. For instance, Halifax party organizer Bill Findlay had ordered Dane Parker to burn one of his books, written by Bukharin. Later party officials learned that Parker, who desperately needed work, had taken a non-union job as a house painter, and they called Parker before a party control commission to determine whether he should be expelled. According to Parker he was cross-examined at a meeting attended by ten or fifteen people, many of them his friends, and then expelled. Parker was readmitted in 1940 after Roscoe apparently intervened on his behalf.

Roscoe thought these heavy-handed actions were ridiculous and frequently criticized the party for such behaviour. But he seemed to think such incidents were not worth resigning over. Then, near the end of the Depression, the troublesome question of what was really going on in Moscow took a back seat to a new concern. The power of Nazi forces in Germany had grown tremendously and Hitler had vowed to wipe out communism wherever he found it. For Roscoe and all other communists, there was a great need to defend the Soviet Union—the world's only socialist country.

FIFTEEN

Party Politics and the War Against Fascism

In 1936 a man living in Sydney named Ronald Gillis, a prolific writer of letters to the papers, wrote to *The Halifax Chronicle* singing the praises of fascist dictator Benito Mussolini. Gillis said that during the ten previous years Mussolini had done more for the Italian people in the way of employment, housing, and economic performance than any other government in Europe had done for its people.[1]

Roscoe Fillmore, an equally prolific writer of letters to the editor, was not one to let this go by without rebuttal. He responded with a harsh condemnation of Italian fascism, arguing that thousands of Italians who objected to the methods of the dictator Mussolini were being held in prison camps.[2] For Roscoe fascism was a declaration of war on all things that civilized people had worked for through the ages. People had fought their way out of slavery and serfdom but, Roscoe feared, fascism had the potential of cancelling out all those gains.

The largest fascist group in Canada was in Quebec, where leader Adrien Arcand claimed a membership of eighty thousand and had formed the National Unity Party. In Halifax William Crane had been organizing fascists since about 1933. The group was small but active. It distributed flyers urging merchants to identify themselves in their advertising as Gentiles so Jewish stores could be boycotted. At least two stores adopted such a policy. Beckie Buhay, who was visiting Halifax, wrote in *The Clarion* that fascists had tried to prevent a Jewish doctor from having operating privileges at a Halifax hospital on the grounds that he would murder Gentile children.[3] There were rumours that members of the Nova Scotia legislature had donated money to the fascists. During the summer of 1939, Adrien Arcand toured several towns in New Brunswick, where one Daniel O'Keefe was working to recruit new members for the movement.[4]

People in Canada seemed slow to react to the fascist threat. In 1936, less than three years before the outbreak of World War II, the Germans were gathering military intelligence in the Maritimes. Canadian authorities gave a detailed tour of the ports of Halifax and Saint John to a German expert on port construction, who was most likely gathering information for the Nazi military. In the same year the German dirigible *Hindenburg*, fully equipped with espionage technology, flew over the Maritimes without permission.[5] In 1937 the Canadian military warmly welcomed a German warship to Halifax with a salute of cannon fire from Citadel Hill. "A distinct feeling of camaraderie was to be noted," *The Halifax Chronicle* reported, "as the uniformed German sailors wandered at will through the city, jostling with the Halifax crowds and seeking their own amusements."[6]

Meanwhile, the news from Europe was not encouraging. Hitler kept demanding territorial concessions, and Britain and France, hoping to avoid war, kept trying to appease the Fuhrer. Roscoe, who had begun to focus his attention on the trouble in Europe, argued that a policy of appeasement would not stop Hitler. He strongly criticized the main author of this policy, British Prime Minister Neville Chamberlain, for allowing Hitler to gain new territories and resources that would allow the Nazis to build a strong war machine.

As it became increasingly clear that war was inevitable, one of Roscoe's greatest fears was that an all-out conflict might threaten the survival of the Soviet Union. Hitler had vowed long before 1939 to destroy communism. Realizing that Russia was not prepared for war, Stalin at first tried to reach an agreement with the Allies and then, not trusting the British and French, pursued simultaneous peace talks with both the Allies and the Germans. In August 1939, when it appeared that the British were not taking their talks with the Soviets seriously, the Soviet leader signed a non-aggression treaty with Germany that declared the neutrality of Russia in the event of war.

The Soviet shift created a new dilemma for Roscoe and the other socialists who supported the Soviet Union. It flew in the face of Roscoe's bitter opposition to fascism; yet, he reasoned, the Soviet Union had to protect itself and had to protect the Communist revolution. Despite the treaty, Roscoe thought that the Russians would still eventually be attacked by Germany.

The developments of the autumn of 1939—after Germany declared war on September 1, Britain and then Canada (a week later) made their declarations of war—left Roscoe in a quandary. The one nation that he felt close to had sided with the fascists in a war against Britain and its allies, including Canada. He still argued that Stalin was doing what was necessary to secure the survival of the country. But he disagreed with the position of the Canadian Communist Party, which said Canadians should not support the war against Germany. The Communists were distributing leaflets opposing the war across the country, including the Maritimes, a matter that aroused further attention from the RCMP.[7]

Apparently, during the twenty-two months of the Russian-German alliance Roscoe did not publicly state his support for the Soviet Union nor did he express his opposition to fascism. He did not write his column for the *Steelworker and Miner* during this period and stopped sending letters to newspapers. He may have chosen silence as a way to avoid being seen as opposing the Soviet Union. He would also have been aware of the unpopularity of his views, and he may have been afraid of losing his job with the DAR or, worse still, of being interned. In September 1939, the government enacted the Defence of Canada Regulations, giving it full power of censorship over the press, the ability to detain people without trial, and the right to detain anyone who made statements that might interfere with the conduct of the war.[8]

As a well-known supporter of the Soviet Union, Roscoe's situation was particularly awkward, even dangerous. People were angry that the Soviets had gone further than just signing a non-aggression treaty with Germany. The Soviet Union provided Germany with huge shipments of raw materials in exchange for military supplies, and at Hitler's invitation Soviet forces had crossed Poland's border and seized control of nearly half the country for itself. Moscow also seized the Baltic states of Estonia, Latvia, and Lithuania, territory that Russia had lost during World War I. Even though the Allies cried "Russian imperialism," Roscoe saw these invasions as necessary for the future defence of the Soviet Union. But his neighbours in the Annapolis Valley, who tended to be extremely patriotic for Britain, hated the Soviet Union for siding with Germany.

The Fillmores were engulfed in a militaristic atmosphere in the Valley as thousands of young men began to enlist. Soldiers for the infantry trained at Camp Aldershot, three kilometres from Centreville, and military manoeuvres were a daily occurrence on the road in front of the Fillmore household.

Sympathizers to the Soviet cause automatically came under suspicion by the Canadian government and the RCMP. The Communist Party paper, *The Clarion*, was suppressed and the party and fourteen "auxiliary" organizations were declared illegal. The RCMP arrested and interned more than one hundred people, and some went underground to avoid internment. Tim

Buck was among a handful of party members who fled to New York. In Halifax, Dane Parker joined the Canadian army against the protests of the Communist Party. Parker was discharged because of his past Communist activities, so he got on a ship heading for England, where he again joined the Canadian army, this time under a false name, and served until the end of the war. Joe Wallace was among those interned. Wallace, one of the main organizers of the Canadian Labor Defence League, was earning a meagre living as a freelance writer for the *Canadian Railroad Employees Monthly* and *The Clarion* in Toronto.[9] Several of the top officials of the Canadian Seamen's Union, including Pat Sullivan, were also interned, as was Charlie Murray of the Canadian Fishermen's and Fish Handlers' Union.

Murray's internment was directly related to his union work, and his arrest caused considerable anxiety for him and his wife Kaye. She was due to give birth to their second child at the very time when the RCMP arrived at their Lower Sackville home in October 1940 to arrest Charlie. Showing no concern for Kaye's condition, the RCMP hurried Murray off to Halifax, leaving her alone in the house. The Mounties wanted to put Murray immediately on a train and send him to an internment camp at Petawawa, Ontario. Murray, accusing them of using Hitler-like tactics, won a small concession: He was allowed to spend the night in the Halifax jail. A daughter, Heather, was born the following day and Murray was taken to the hospital. "I was permitted to see my wife for a few minutes," Murray later wrote, "but just in the presence of the officers, never alone." That evening he was taken to the train and escorted to Petawawa. During the trip Murray's RCMP escort sat in civilian clothes with the butt of his gun sticking out of his shoulder holster. The escort kept telling Murray, "Oh, how I would like you to make a break for it so I could shoot you down, you red bastard!"[10]

At Petawawa, Murray and other Communists were forced to share living quarters with fascists, and he and the other Reds feared for their safety. Later Murray recalled that during a celebration commemorating an Italian holiday, a throne was built for Canadian fascist leader Adrien Arcand. "He was formally seated on the throne and he announced the names of the fascist cabinet who were to accompany him on his march to Ottawa as soon as Hitler won the war."[11] Murray and other communists were later moved to the Hull jail, which sat on the bank of the Ottawa River within sight of the Parliament Buildings. Joe Wallace, who was interned at Petawawa and in the Hull jail, said he was sent to solitary confinement for a month for correcting an officer who had called him a Russian. Asked what he did for the four weeks, Wallace replied, "I wrote poetry on the wall with a nail."[12]

During Murray's internment, Kaye struggled to provide enough food to feed her two daughters. For a while the family was on welfare. Roscoe, who had become a close friend of the Murrays, occasionally visited Kaye, encouraging her and providing gifts for the children. When the Soviet Union entered the war on the side of the Allies and Charlie and the other communists were still not set free, Kaye went to Ottawa as part of a delegation of wives to demand their release. After a total of eighteen months' internment, Murray, Wallace, and most of the other Reds were released.

❧

Roscoe was not interned, but the Fillmore home in Centreville was raided. In early 1940 a team of several RCMP officers and members of the Military Police from the base at Aldershot arrived at the home in the middle of the night. When Roscoe came to the door they said they were there to conduct a thorough search and proceeded to go through the house from top to bottom. Fifteen-year-old Frank was understandably frightened to be awakened by the police entering his room. The police went through all of his pen-pal letters, which he had been exchanging with other teenagers.

The RCMP were searching for subversive literature, but fortunately the material they were looking for had already been removed from the house. A few weeks before the raid, bundles of Communist materials that Roscoe had not wanted arrived for him at the train station in Centreville. The bundles probably included illegal pamphlets that denounced Canada's participation in the war. Roscoe and Frank had boxed up and buried the materials, along with copies of *The Communist Manifesto*, under a spruce hedge that separated the Fillmore property from the MacDonald's next door.

After the raid Margaret wrote J. Lorimer Ilsley, their Member of Parliament and the Minister of National Revenue in the Liberal government, to protest the police action. Ilsley wrote back saying Roscoe wouldn't be bothered again, a gesture in sharp contrast to something he had done just a few months earlier. Soon after the outbreak of the war Ilsley sent a letter to RCMP headquarters informing them that "loyal constituents" in his Annapolis Valley riding had provided his office with the names of three people—Fillmore, James Sim, and Charlie MacDonald—who were regarded as Communist agitators.[13]

Even though Roscoe was not publicly expressing his views about the war and the Soviet Union, security agencies kept a close watch on him. Letters he sent to the United States were intercepted by the Canadian Postal Censorship office, which gave copies to the RCMP, External Affairs, and National

Defence.[14] Roscoe was concerned that one letter, written to a New York radio talk show host, contained comments that might have been considered treasonous because he questioned Britain's commitment to the war.

Letters that Roscoe wrote to his daughter Ruth in Detroit were also seized. Roscoe's half-sisters Mabel and Clara were concerned about their connection with Roscoe because their husbands were working on the production of materials for the war. Willard, living in Detroit, also was writing provocative, controversial letters concerning the war to Roscoe.

❧

Relations between Berlin and Moscow had grown tense. Hitler was angered by how quickly Stalin had seized the Baltic states and a part of Rumania. In June 1941 Germany invaded Russia, facing little resistance as their tanks quickly penetrated and captured the Ukraine. Hitler's troops advanced to within twenty-five kilometres of Moscow, before near-constant rain bogged down the tanks in heavy mud. Then came the winter. German troops were unprepared for the freezing cold. The Russians counterattacked and inflicted heavy losses on the retreating enemy.

The German invasion of the Soviet Union led to dramatic changes, both in the direction of the war and for Roscoe personally. As Roscoe had suspected, Hitler had declared an all-out war on communism. Hitler told the chiefs of his armed services that all communist leaders were to be killed: "This struggle is one of ideologies and racial differences and will have to be conducted with unprecedented, unmerciful and unrelenting harshness."[15] When Britain and the Soviet Union signed a peace treaty the Russians suddenly became allies of Canada, and Roscoe was now overwhelmed by the interest people showed in the Soviet Union. Neighbours would stop him to talk about the Russian war effort.

Roscoe began to speak out strongly in support of the Soviet Union, writing angry antifascist columns in the *Steelworker and Miner* and sending a flurry of letters to the local newspapers. He was the central figure responsible for the establishment of a local ad hoc group opposing fascism. The group had got together when Saskatchewan MP Dorise Nielsen came to the province and spoke about fascism to a group of people in the school house at Woodville, a village not far from Centreville.[16] After Japan entered the war Roscoe drove his gardening truck around the Annapolis Valley with a sign on the side that declared, "Don't buy German Goods. Don't buy Japanese Goods!"[17]

A month after Russia's entry into the war, Roscoe wrote a letter to *The Halifax Chronicle* saying it was scandalous that Canada allowed the governments of three enemies—Vichy France, Finland, and Japan—to have full diplomatic status, including the privilege of sending secret messages out of the country, while Canada's new ally, the Soviet Union, had been without diplomatic representation in Canada for years. The Soviets had established offices in Canada in 1942. Roscoe had become friends with the Soviet vice-consul in Halifax, Mikhail Kutsenko and his family, frequently visiting them when he went to the city.[18] He helped to arrange a tour of farms in the Kentville area for Kutsenko, and the vice-consul announced that the fruit dehydration technology he was shown could help to overcome famine in Europe once the war was over.[19]

Roscoe was concerned that the Allied forces weren't doing enough to help the Soviets. Stalin wanted the Allies to open a front against the Germans in northern France to take pressure off the Russian front, which was bearing the brunt of the German offensive. Roscoe wrote to *The Halifax Chronicle* saying that the thousands of young Russians who were being killed on the Eastern front were saving the lives of an equal number of Canadian young men. When the Germans launched another attack on Russia in the summer of 1942, penetrating close to the valuable oil fields in the south, Roscoe complained that while Russia was being battered, the British, U.S., and Canadian forces had an active force of more than four million men who were doing very little to win the war.

Roscoe shared Stalin's suspicions that the Allies were leaving the fighting for as long as possible to the Soviets. Churchill's hatred of communism was well known. Harry Truman, who succeeded Roosevelt as president in the late stages of the war, seemed to want to ensure that the Soviet Union emerged as an exhausted victor.[20] Roscoe had both political and humanitarian concerns for the Soviets. The Allied leaders knew that Soviet and other Eastern European citizens were suffering terribly at the hands of the Germans. More than seven hundred thousand Jews and Slavs died in mass executions in Russia. In prisoner-of-war camps, all Jews and Communist functionaries were shot. During the course of the war more than three million Russian prisoners of war were starved to death or died from the harsh weather. When Hitler realized that the war was not going to end as quickly as he had once thought, he had three-quarters of a million Soviet prisoners of war taken to Germany to work as slave labour in factories producing weapons.[21]

After the disastrous Dieppe raid in the summer of 1942, Roscoe wrote in *The Halifax Chronicle* that what was needed were raids of one hundred times the scale of Dieppe. He wrote that to stage a solitary Dieppe raid—despite its four thousand casualties—awakened hope that a major invasion was planned, only to have that hope turn to frustration and despair when no main invasion followed. Roscoe's criticisms angered Nova Scotia attorney-general Lauchie Currie, who happened to be giving a speech in Kentville the same week that Roscoe's letter appeared in the paper. Currie ridiculed Roscoe for advocating the needless death of many soldiers.

Roscoe defended his right to question the progress of the war. "Our sons and daughters have gone into the forces," he wrote in *The Halifax Chronicle*. "We, the middle-aged and old, spend our nights worrying and wondering what is in store for us, and the days in dread lest the mail man bring us bad news. So we have every right in the world to be interested, to criticize, and we expect that right without apology."[22]

His words had special meaning for Roscoe and Margaret because their two youngest children had signed up for duty with the Canadian forces. Both Allie and Frank shared Roscoe's concern about the threat of fascism. In early 1942 Allie quit her job as a clerk in a Kentville department store and joined the Canadian Women's Army Corps (CWACS), created to bring women into non-combat jobs and to release men for combat duty. In the fall she was sent overseas to London, where she worked as a chief administrative clerk during the blitz. Frank entered the Royal Canadian Air Force six days after his eighteenth birthday. After blacking out during a training flight at Lachine, Quebec, Frank was transferred to ground duty and served at air bases in Nova Scotia.

With no children left at home, and very little business at the nursery because of the war, Roscoe and Margaret closed up shop for much of the time and moved into Kentville to stay with Rosa and her family. It also was much easier for Roscoe to get to his job with the DAR from Kentville.

Roscoe continued to press his concerns about the Canadian aspects of the war. He wrote to *The Halifax Chronicle* criticizing the Canadian establishment for pushing Quebec on the issue of conscription and thus popularizing the small fascist movement in the province. In another letter he wrote that if the people of Quebec were lacking in enthusiasm for what they perceived to be a "foreign war," they should be told about the Nazi U-Boat excursions up the St. Lawrence River. Roscoe said that instead of taking the opportunity to show people how real the war was, the incident was hushed up. The existence of the U-boats became public knowledge only when a Quebec MP talked about their presence. He was scolded by the naval minister, who claimed the release of the information by the MP would aid the enemy.

In yet another letter, in May 1942, Roscoe said that the Canadian government and the RCMP were not concerned about the growth of fascism at home. He said that many fascist plots were being uncovered in the United States, and he wondered why there weren't arrests of fascists in Canada. He believed that fascists had been behind many of the anticonscription riots and demonstrations in Quebec. Roscoe thought that if Communists had been involved in such demonstrations, they would have been beaten and sent to internment camps without a hearing. He was not far off the mark. In 1941 RCMP Commissioner S.T. Wood wrote an article, "Tools for Treachery," saying that Communists, not Nazis or fascists, constituted the force's most troublesome problem in Canada. The official historian of the RCMP later wrote that Communists were the "principal target of RCMP intelligence" during the war.[23]

By 1943 the important role that the Soviet Union was playing in defeating the Germans had led to a further change of attitude in some quarters of the conservative Annapolis Valley. Roscoe almost certainly influenced the Kentville *Advertiser* to carry an editorial that praised the "valiant efforts and heroic sacrifice" of the Soviet people and urged Nova Scotians to make a donation to the Russian Relief Fund.[24] In April 1943 Roscoe was invited to speak to the Kentville Rotary Club about his experiences in Siberia and the Soviet Union's role in the war. He didn't mince words, telling the business-oriented gathering that Canadians were in part responsible for the war because they had ignored the growth of fascism in Germany during the 1930s.[25]

❧

As it became apparent that the Allies would win the war, Roscoe and other communists began to think about the kind of society they wanted to see in Canada after the war—and about working to avoid the unemployment and economic chaos that had existed during the Depression. It seemed possible that Canadians would become more radical in their political outlook and accept a more radical leadership. Roscoe reasoned that because the Canadian government had spent billions of dollars on the war effort, it wouldn't be so easy to argue that there was no money to provide jobs, decent housing, health care, and other social benefits. He hoped that communism might not be the bogeyman it had seemed at the start of the war. The hard-fought victory by the Russians at Stalingrad had won over the hearts of many Canadians. Because of pressure from the Allies, Stalin abolished the Comintern in 1943, which was meant to be a signal that the Soviet Union was no longer intent on spreading revolution to the Allied countries.

A campaign to legalize the Communist Party was stepped up. Tim Buck travelled across the country in late 1942 to pressure the government. Despite the campaign, justice minister Louis St. Laurent refused to lift the ban, declaring that the fundamental philosophy of the communists was in conflict with his conception of Christian civilization.[26] When it appeared that St. Laurent wouldn't act before the next election, the communists moved to establish a new political party, quite sure that with their new popularity the government wouldn't shut them down.

In August 1943 Roscoe travelled to Toronto to attend the founding convention of a political party that would be, in every way other than name, a communist party. In the weeks before the convention, three communists were elected in Canada. Two of them had run as Labor candidates in Ontario and the third, Fred Rose, had been elected to Parliament in a by-election in Montreal. There was a tremendous sense of optimism as the convention opened at the King Edward Hotel. Roscoe was impressed by the dedication and hard work shown by the more than five hundred delegates. Their energy and enthusiasm was a tonic for him. He was pleased by the large turnout of delegates from Quebec and by the eloquent and hard-hitting speech of Dorise Nielsen, who had been elected an independent member of the legislature in Saskatchewan but announced at the convention that she was joining the new party.

In the excitement of the convention, when the chairman asked for proposals for the name of the new party, Roscoe shouted from the back of the room: "How about the Labor-Progressive Party!" The name was accepted and thus, according to Tim Buck, Roscoe had named Canada's new political party.[27] Roscoe was elected a member of the central committee of the party, representing mainland Nova Scotia. Fred Brodie represented Cape Breton and Rev. John Mortimer represented New Brunswick.

The program adopted by the Labor-Progressive Party (LPP) was different in many respects from its predecessor, the Communist Party of Canada. There were no references in the party's program to Marxism-Leninism, and only one fleeting allusion to Marxism. The program mixed the immediate goal of defeating fascism with its hopes for the postwar world. The goal of forming a socialist state was left to the distant future when the majority of people would be ready for such a step.[28] Roscoe was particularly pleased with the LPP's position toward social democrats, such as the CCF. Years earlier the Communist Party had bitterly attacked social democrats—a policy that Roscoe had opposed. Now the new party campaigned to join with the Liberals and the socialists to attempt to keep the Conservatives out of power. Shortly after the convention Roscoe wrote that while the LPP had its own objectives and point of view, it would work "in the most complete harmony with the CCF wherever and whenever opportunity allows."[29]

Back home in Centreville, Roscoe immediately began working to organize a Nova Scotia branch of the LPP. When the party held its founding convention in New Glasgow in December 1943 he gave the opening address. It was an emotional talk about the historic significance of the Toronto meeting and about how after more than thirty-five years' experience in socialist activities in Canada, he now believed there was considerable public acceptance of the idea of radical socialism.

Labor-Progressive Party organizers were not particularly discouraged that only thirty-eight people attended the New Glasgow convention. There was a small, solid core of people from many areas of the province, including Charlie Murray and Scotty Munro from Halifax, Roscoe and Jim Sim from the Valley, Fred Brodie and Wesley Bond from Cape Breton, and George MacEachern and Sarah and John MacKay from Pictou County. Also present were John Mortimer and two other observers from New Brunswick. MacEachern was elected party chairman and Fred Brodie provincial organizer.[30]

Along with travelling to different areas of the province to promote the LPP, Roscoe started to write a research report that would form the basis of the party's economic strategy for the region. According to Vern Bigelow, a young LPP member who came from a conservative background in Canning, Nova Scotia, Roscoe was an effective spokesman. "People just flocked to him; they were drawn to him," Bigelow said. "He was speaking to people who didn't know the basic theories, and he was talking about very basic old socialist things."[31] Roscoe had the ability to translate abstract concepts into terms that could be easily understood. Another admirer of Roscoe's was Fran Fassett, who married James Sim Jr. She said Roscoe "had time even for the minions," meaning people like herself who were new to socialism.[32] Young people such as Bigelow and Fassett delighted in Roscoe's company and admired his sense of humour in difficult times.

Jimmy Sim, Roscoe's close friend, was also active in the LPP as a member of the party's provincial executive. The two men often got in Roscoe's old half-ton truck and headed for political meetings in Kentville or Halifax, sometimes accompanied by Margaret, who would go shopping and attend the social part of LPP gatherings.

Sim was not a strong speaker, but he was a steady and hard worker for the socialist movement. After Sim's election to the LPP executive, RCMP Corporal E. Swailes of the Kentville detachment commented in Sim's file that he was usually very quiet and did not make a habit of forcing his views.

Noting that Sim was on intimate terms with Roscoe, the report added: "The belief is that, unless influenced by Fillmore, this subject, although sharing the same political views, would not have taken the more active part, which he is now displaying by his election to responsible posts in the Labor-Progressive Party."[33]

Roscoe shared the stage at two LPP meetings in Halifax with Annie Buller, the long-time communist who had been interned during the early part of the war. In many ways Roscoe had become the patriarch of the small Maritime communist movement. He gave uplifting speeches about the history of socialism and his own role in the movement. He talked about his early organizing efforts for the Socialist Party, his work in the Soviet Union, and how the international communist movement had helped defeat fascism.

❧

Roscoe was encouraged enough by the success of the LPP in the winter of 1943 that he held meetings with his socialist friends in the Valley and decided to run for office in the Digby-Annapolis-Kings riding in the next federal election, expected for 1945. Roscoe knew from the outset that he had no chance of winning, that he'd pick up, at most, a few hundred votes. As a candidate advocating radical change, Roscoe was at even more of a disadvantage than usual because the Valley was prospering during the war. He wanted to run mainly to acquire a forum for his views in the new environment after the war.

One of his opponents would be lawyer J. Lorimer Ilsley, the Liberal government's finance minister and a Member of Parliament since 1926. The Liberal government of Mackenzie King, with its large wartime majority in Parliament, was expected to be a repeat winner in the next election. Ilsley used his position as an influential cabinet minister to funnel federal money into his valley riding at an unprecedented rate, making the Valley more prosperous than at any other time in history. The army base at Aldershot was expanded, a new air-force base was located at Greenwood, and a new naval base was established at Cornwallis.[34] Valley businesses were told that they should do what was necessary to make as much profit as possible from the markets provided by the war effort. There were reports of local merchants driving into Aldershot and selling a truckload of goods to the army, driving out with the goods still aboard, and returning to sell the same lot again.[35]

Early in January 1944, after the RCMP had heard of Roscoe's decision to run, A.N. Eames, commander of the Halifax division, wrote a critical, two-page report for head office. Eames stated: "Roscoe Fillmore is not qualified to run as a representative of the people. He is a Marxist with only one-sided reasoning and no regard for human thought or feeling unless it is his own brand."[36]

Roscoe's campaign committee—headed by Jimmy Sim along with W.C. Lockhart, a locomotive engineer with the DAR, and Clarence Coffill, a Kings County farmer—decided to run Roscoe under the Farmer-Labor Party, despite his membership in the new Labor-Progressive Party. Communists in Canada had often referred to themselves as a farmer-labour party, and Roscoe's election committee most likely felt that the Farmer-Labor label would be better suited to the Valley constituency than Labor-Progressive. Still, Roscoe told at least one meeting that there was little difference between his policies and those of the LPP.[37]

The group opened a campaign office in Kentville and planned public meetings for the first and third Sundays of each month. LPP provincial organizer Fred Brodie of Glace Bay came to the Valley to help Roscoe in the election.[38] Beckie Buhay and Annie Buller, the two most prominent women in the Canadian communist movement, also campaigned on Roscoe's behalf for a few days. Roscoe's son Frank helped when he was on leave from the air force. Frank and his wife, Irene, became members of the LPP, making Frank the only one of Roscoe's children to hold membership in a communist party.

Roscoe's election campaign was considerably toned down in comparison to his radical positions of earlier days with the Socialist Party. He emphasized a commitment to continuing the war against fascism, the punishment of all Nazi war criminals, and the exposure and destruction of all pro-fascist forces in Canada. Locally the key for Roscoe was a promise to provide assistance to the troubled apple industry. Sim figured Roscoe's best chance of making a strong showing would be to become "champion of the apple growers."[39] There were more than 2,500 small commercial apple growers and their industry was in an upheaval. British markets had been lost at the beginning of the war, inflation was taking its toll on farmers' incomes, and money lenders and speculators were preying on the hard-hit farmers. Roscoe said that many farmers were being forced to seek high-interest loans from private individuals because the chartered banks refused to give them credit.

To get the supplies necessary to operate for the coming year, many farmers were borrowing against the revenues they would receive from their next crop. They were signing notes pledging to pay up to 12 per cent interest—much higher than the interest on a bank loan. They were in danger of losing their farms altogether or—as Roscoe put it—of becoming mere foremen on their own farms. Roscoe pointed out that one loan shark alone had taken control of 169 farms in the Valley through extending personal, high-interest loans.[40] He was apparently referring to George Chase, a Valley businessman who controlled the shipment of much of the apple crop to the British market and was foreclosing on many apple growers who had overextended their line of credit.[41]

Roscoe promised a farm policy that included low interest rates for farmers caught in the grip of the money-lenders and speculators, as well as crop insurance and cold storage plants in which apples could be kept until there was a demand.[42] He said the farming community and small business needed protection from the conglomerates that had grown so big and powerful during the war.[43]

Other aspects of his platform included a planned transition of industrial plants and workers from wartime to civilian work, without shutdowns and layoffs; immediate establishment of new rural and urban housing; an all-embracing system of medical and hospital services; and free education for all children.

Roscoe had to do much of his campaigning in the evenings or on weekends because of his job with the DAR. Still, he managed to travel the length of the riding, giving dozens of speeches. At the Cornwallis Inn in Kentville he attacked the government for selling war materials to Germany before the war, for failing to institute complete conscription, and for allowing fascism to grow in Canada. He also spoke in small fishing and farming communities such as Morden, Scott's Bay, and Baxter's Harbour. Even the RCMP seemed impressed by Roscoe's speech at Baxter's Harbour. The Mounties duly reported, "The speech was well-received by quite a number of people attending, particularly amongst the poorer classes, and evidently Fillmore has a command of his subject which impresses his listeners."[44] Meetings in Wolfville and Hantsport, though, were poorly attended.

Roscoe also used the medium of radio to reach voters, delivering eighteen radio broadcasts during the campaign. Most of his speeches were heard on CHNS in Halifax, but he travelled to Charlottetown and purchased fifteen minutes of time on radio station CFCY for a political broadcast that was heard through most of the Maritimes. He told listeners that he was running for office in the hope of representing the 80 per cent of Canadians —farm families and wage earners—who were badly underrepresented in government.[45]

When the date of the election—June 11, 1945—was announced, Roscoe had already been campaigning for eighteen months. He hoped that the CCF, which had never run a candidate in Digby-Annapolis-Kings, would not put anyone forward this time. With no CCF candidate Roscoe would pick up all of the protest vote. But the CCF was emerging as a strong political force in much of the country. The party formed the official opposition in British Columbia and Ontario and, in 1941, had elected three members in Nova Scotia to become the official opposition. In 1943 the CCF topped a national Gallup opinion poll, and in 1944 the first ever CCF government in Canada was elected in Saskatchewan.

The CCF was hoping for a major breakthrough and, with only six weeks left before election day, the party nominated a candidate to run in Roscoe's riding. Roscoe was furious, having been assured by rank-and-file CCFers that there would be no CCF candidate. Some CCFers had even contributed to Roscoe's election fund. The party nominated Cecil Hansford, a barber from Wolfville who was in the armed forces and stationed at Camp Aldershot. Roscoe, who knew every socialist in the Valley and had never heard of Hansford, was surprised to read in the paper that Hansford was being introduced at his nominating convention as an active socialist and a great fighter for the working people.

In Roscoe's mind the CCF candidacy in his riding was a serious mistake for a party supposedly concerned about the best interests of the working class. Roscoe supported the LPP policy that the only way the Conservatives could be defeated was through a united front of Liberals, socialists, and communists. The CCF, however, proceeded to run candidates in 206 of the 245 total ridings, its greatest number ever in a federal election, and in almost all of the sixty-eight ridings contested by LPP candidates. Roscoe believed that the working-class vote was being divided and wasted, largely through the policies of the CCF.

Roscoe called a meeting of his election committee and offered to withdraw his candidacy if his group felt that was the best thing to do. The committee declined the offer and Roscoe continued to campaign. The final few weeks of the campaign took place amid the jubilation and chaos of VE-Day. In Halifax three people died in riots after the celebrations turned into a wild drinking and looting spree.

On election day Roscoe received only 362 votes. The results seemed to indicate that while the Soviet Union had become respectable, it still was not respectable to vote communist. The CCF candidate received more than double Roscoe's total. Ilsley was easily re-elected, and the King government was returned with a reduced majority. The most votes received by any of the three LPP candidates in Nova Scotia were James Madden's 920 in Cape Breton South, where Clarie Gillis won the CCF's only seat east of Ontario.[46]

The only Labor-Progressive candidate elected in the country was Fred Rose, who won in Montreal-Cartier where there was no CCF candidate. The CCF won a total of twenty-four seats across the country. LPP leader

Tim Buck would most likely have been elected in Toronto-Trinity if there had not been a CCF candidate. As it was, the riding was won by a Conservative.

Despite Roscoe's poor showing, the Farmer-Labor Party continued to hold twice-monthly meetings in Kentville. Roscoe also continued to be active in the Labor-Progressive Party. During and after the election, he had several long talks with Ilsley, who suggested that Roscoe would have an excellent future in politics if he joined one of the mainstream parties—a suggestion that Roscoe was not likely to take seriously.

❧

During the week before the election, the war in Europe had ended. Roscoe and Margaret were thankful their two children were unharmed. Frank had spent his three years in the forces at bases in Nova Scotia. He had been slated to go to the Far East in 1945 with a special RCAF group, but the trip was cancelled when the atomic bomb was dropped on Hiroshima.[47] Allie, promoted to platoon officer and quartermaster in the CWACS in Europe, went to Holland in July 1945 to help that country recover from the war. She served thirty-seven months overseas and was awarded the Defence Medal, returning to Canada in late 1945 with the rank of lieutenant.

The defeat of Nazi Germany was a momentous occasion in Roscoe's life. He hated fascism and for the rest of his life frequently referred back to the threat of the Nazis and how they had been defeated. In Germany the Allied and Russian troops soon uncovered the horror of the Nazi death camps, dulling the sense of triumph. Shortly after the war-crimes tribunal at Nuremberg announced its decision—to execute twelve top leaders of the Third Reich and send another seven to jail—Roscoe expressed a strong sense of dismay in the pages of the *Steelworker and Miner* that the majority of Nazis were going unpunished.[48] By the end of 1946 the United States alone had granted amnesty to eight hundred thousand Nazis in the U.S. zone of Germany. Many of these people, who included scientists who had helped develop the Nazi war machine, ended up in the United States. As Europe began to rebuild, Roscoe and others warily watched the emerging ambitions of the United States unfold and a new political climate—the Cold War—set in.

SIXTEEN

Cold War Days: The Decline of the Labor-Progressives

MUCH OF THE enthusiasm Roscoe Fillmore had for the future of the Labor-Progressive Party in Canada stemmed from his belief that there would be a general liberalization in the Soviet Union immediately after the war. He reasoned that radical socialist parties in countries such as Canada would reap the rewards of this liberalization.

But on a Saturday morning in February 1946, Roscoe woke up to read a startling headline in *The Halifax Herald*: "Canadian Government Employees are Suspects in Espionage Ring." Back in September, Igor Gouzenko, a Soviet intelligence cipher clerk wishing to defect to Canada, walked out of the Soviet Embassy with evidence that the Russians were spying on Canada and the United States. Gouzenko knew there was even a spy working in a Canadian research lab where atomic experiments were going on. When the story was finally told to the Canadian public four months later, people were shocked and couldn't believe that Russia, the wartime ally, would spy on Canada.

Perhaps attempting to play down the significance of the Gouzenko revelations, Roscoe wrote in the *Steelworker and Miner* that he was not particularly surprised: Such activities were only natural among countries. But the Gouzenko affair was a major factor in the buildup toward the Cold War. Thirteen people were arrested, including Montreal LPP MP Fred Rose, who was convicted of recruiting for Soviet intelligence and sentenced to six years in prison.[1]

Three weeks after the Gouzenko scandal broke, Winston Churchill delivered his anticommunist speech at Fulton, Missouri, calling for war preparations against the USSR. The Cold War was on. Roscoe had hoped that a warming of Western powers toward the Soviet Union would persuade Joseph Stalin to ease censorship and individual freedoms at home and lessen military tensions abroad. If there had been an opportunity for the Soviet Union to show a new face to the world, it disappeared amidst the attacks of Churchill and U.S. president Harry Truman.

The Soviet Union was reeling from the war, its economy in tatters. At least twenty-five million Soviets had died during the war and another twenty-five million were homeless. In an attempt to rebuild the country Stalin launched an ambitious five-year plan and ruthlessly implemented it through the mechanisms of the Communist Party. The strict regimentation was carried over into literature, the arts, and the sciences. Recalling the tremendous war effort that Stalin had led against fascism, Roscoe thought that history might one day judge the Soviet leader as one of the greatest of men. But Roscoe strongly disagreed with Stalin's policies after the war. "He missed the psychological moment when a great gesture and a slackening of censorship, secrecy and suspicion might have won half a world," Roscoe later wrote. "The mistakes most of us make have no such dire effects, but the mistakes of a Stalin can be world-shattering."

With a steadily rising anti-Communist hysteria evident in both Canada and the United States, Roscoe feared that the presence of spies in Canada would be used to enable the forces of big business and fascism to attack first the communist and labour movements and then the USSR. He grew uneasy as many people expressed the view that a war should be carried out against the Soviet Union before it developed the atomic bomb for itself. Even with his apprehensions about Stalin, there was no doubt in Roscoe's mind about the significance of a possible Soviet defeat. "It would be the most grievous and serious setback civilization has experienced in all history," he wrote in the *Steelworker and Miner*.[2]

Roscoe continued to devote considerable time to radical politics even though he was increasingly busy. At the end of the war his son Frank, at twenty-two, was keen to re-establish the gardening business, and the two men became partners in the Valley Nurseries. In 1945 Frank, his wife Irene, and I—age two—moved into the Fillmore home at the nursery in Centreville. The next year we were joined by Roscoe and Margaret, who had lived in Kentville during most of the war. Roscoe kept his job as head gardener with the Dominion Atlantic Railways and also took on the job of looking after plant propagation for the nursery. Frank took on the bulk of the work associated with running what would soon be a rapidly growing business.

Despite his professional commitments, Roscoe still found time to give to the Labor-Progressive Party. But there soon would be dissension within the ranks of the Nova Scotia party. Near the end of the war, Ethel Meade, an American who had belonged to the Communist Party in Pittsburgh, arrived in Halifax with her husband, Bert Meade, who was port agent with the Canadian Seamen's Union. Ethel, who seemed to be well connected to Communist Party headquarters in Toronto, soon replaced Charlie Murray as party secretary, a non-paying position. She was firmly in control of the apparatus of the Nova Scotia LPP during the months leading up to the federal election of 1945. She operated the party out of the office of the Seamen's Union, where she was employed as a secretary.

But Meade quickly became a controversial figure. Several party members, including Roscoe Fillmore, Jimmy Sim, and Charlie Murray, began to criticize Meade's work as party secretary. Roscoe felt that the party reacted badly to criticism and that there was too much secrecy. Murray said Meade operated the party with the support of a small clique of Halifax supporters. "Ethel wasn't the kind of person who could get co-operation," Charlie Murray said later. "She didn't believe in co-operation. She believed in dominance, and the party was a volunteer organization."[3]

Roscoe's dissatisfaction extended beyond his frustrations with Ethel Meade. He was unhappy about the way both the Canadian and international communist movements were taking a hard line against the leader of the U.S. party, Earl Browder, who advocated co-existence between the communist and capitalist forces of the world. The Stalinist view, which he felt was the view held by Ethel Meade, was that class conflict between the two forces was inevitable and that communists must be prepared for war. Roscoe appeared to have changed his position on how rigid an approach the movement should take. In earlier years, during the crisis in the Socialist Party in 1911 and throughout World War I, he supported expulsions to keep the movement pure. Now Roscoe disliked the rigid position taken by the communist establishment and the manner in which Browder was driven from the movement. He thought that such a rigid position was destructive.

For her part, Ethel Meade said that the members' questioning of LPP procedures was a sign that the party was expanding and departing from the conspiratorial nature prevalent a few years earlier.[4] She also felt there was a parochial nature to the Maritime community. "I never felt like anything but an outsider there," she said later. "If you came from as far away as Sydney you were an outsider in Halifax. It was a very closed community. I felt most of the differences were between the Maritimers and those from away."[5]

Questions about Meade's intentions continued to be raised. Charlie

Murray later complained that Meade made no effort to have the party support his election campaign in Halifax. (Meade's response to this later was that she felt Murray had become soured and disillusioned.) All three of the Nova Scotia LPP candidates had not fared well in the federal election. James Madden received 854 votes in Cape Breton South—the same riding where J.B. McLachlan had gotten 5,100 votes in 1935. Neither Charlie Murray in Halifax nor George MacEachern in Pictou got more than 400 votes. Roscoe, running under the Farmer-Labor label, but really a Labor-Progressive, got 362 votes in the Valley. There were defections from the LPP, and the party was rife with rumours about traitors in the ranks. Questions were raised about the intentions of the Meades. Kaye Murray went to Toronto to ask that Ethel Meade be removed from her post, but no change was made. Later, in 1949, the Meades left Halifax to go to Toronto and soon dropped out of the communist movement.

❧

At first the Cold War did not seem to have much impact on the Nova Scotia LPP. Some fifty delegates attended the 1946 annual convention in Halifax, and in December more than two hundred people turned out to hear Tim Buck speak. The party was generating coverage in the media and there was talk about Roscoe and other candidates running in the next Nova Scotia election. Roscoe was reelected first vice-chairman of the Nova Scotia LPP in 1946 and was responsible for developing the circulation of *The Canadian Tribune* in the Maritimes.

By 1947, however, communists and their sympathizers were being persecuted throughout Canada and the United States for their political beliefs. Even if Ethel Meade had done a poor job of running the LPP, the tremendous pressure exerted on people to abandon the party because of the near-hysterical anti-Red campaign was even more damaging.

Both Charlie Murray and Annie Buller warned the party about spies in their midst, and they were right. RCMP surveillance of the Labor-Progressive Party was relentless following the war. Roscoe's RCMP files show that the force's security division had infiltrated the LPP in both Halifax and Cape Breton. Through sources close to—or inside—the party the police obtained information on the contents of memos sent to party members and private correspondence among LPP members. When Tim Buck made his two speeches in Halifax in December 1946, police came away from the meetings with verbatim accounts of every comment made in the meeting and an analysis of the various groups represented in the audience: "200 present—25 women, 175 men. Approx. 25 Canadian Seamen's Union members. Approx. 20 Dalhousie University students. A group of young Jewish men (Approx. 25). Only a few people over 60, mostly middle aged to 35 made up of persons unknown to me. Meeting well received and loud applause given to many of the remarks."[6]

The attacks against Maritime radical socialists carried over into their work in the Canadian Fishermen's and Fish Handlers' Union (CFFHU) and the Canadian Seamen's Union (CSU). After the war the fishermen's union clashed with a huge new fish company. Financier Ralph Bell had masterminded a business deal to bring together all of the major fishing companies to form one giant firm, National Sea Products. In 1947 the CFFHU went on strike against National Sea and other fish companies to attempt to win a slightly larger share of the profits for the men who worked on the fishing boats. Ralph Bell and other company officials launched a vitriolic anti-communist campaign against the union that has been described as one of the most vigorous anti-union campaigns in Canadian labour history.[7] The company financed a series of publications with titles such as "Communism v. Free Enterprise in the Fishing Industry" and excerpts from sermons called "Christianity or Communism."

Stirred up by this issue, Roscoe went on the attack, writing a letter to the editor that *The Halifax Chronicle* refused to publish, though the *Steelworker and Miner* allowed him to include the comments in his column. Roscoe wrote that while he had thought Nazi propaganda minister Joseph Goebbels had perished in the wreckage of Berlin, he must have survived, "for none other than he could have concocted the pot-pourri of half-truths, red baiting and appeals to prejudice being published in the name of Mr. Bell and his monopoly in fish!"[8] Roscoe charged that Bell's crusade against the alleged Communist leadership of the fishermen was meant to camouflage the damage that Bell and his associates were doing to the fishing industry. A few weeks later Roscoe wrote that the fish monopoly was meant to sew up the fishing industry in the purse of Bell and his associates, "just as Dosco has for years held and exploited coal and steel in Nova Scotia."[9] He wrote that Angus Macdonald and Lauchie Currie looked after "the legal angles" for these monopolies, while the workers quietly starved.

The dispute between the union and National Sea Products was settled when the Supreme Court of Nova Scotia restated the long-held anti-union position that fishermen were not employees of the company and were really co-adventurers who shared in the rewards of each fishing trip. As such, according to the Court, the fishermen did not have the right to strike.[10] The fishermen had been on strike for four months when, once again, the courts had decided against them.

As part of the anti-Red campaign the governments of Canada and the United States vowed to destroy the Canadian Seamen's Union. Hal Banks, a convicted U.S. gangster, came into Canada under the auspices of the Seafarers' International Union (SIU). The CSU had refused to sign a contract offer that would not protect its members from losing their jobs; Banks' union had no such reservations. The SIU signed contracts with ships that were picketed by striking CSU members and, using goons to beat up their rivals, raided CSU vessels.

In Halifax CSU members who belonged to the Labor-Progressive Party were picketing the CNR's ship *Lady Rodney* as part of the strike. Scotty Munro, who had worked with Dane Parker a dozen years earlier to try to organize fish-plant workers on the Halifax waterfront, was the CSU's ship's delegate on the *Lady Rodney*. One night a train pulling freight cars arrived at dockside, and much to Munro's surprise about two hundred SIU scabs, goons, and CN police jumped out and dashed to the ship, scattering the picketers. Taking control of the ship, the scabs—some of them waving sawed-off shotguns—jeered the strikers. When reinforcements for the strikers arrived there was gunfire from the bow of the ship. One seaman was struck in the stomach, another in the eye. After the shooting ended, eight seamen lay wounded on the dock.[11]

The next morning four thousand workers from all over Halifax marched on city hall, demanding to see the mayor. After hearing from Jimmy Bell, head of the Marine Workers Federation, the crowd dispersed. The men who shot the strikers were apprehended, tried, and sent to jail. A few weeks later, when the Trades and Labor Congress abandoned the CSU in favour of the SIU, Roscoe sent TLC president Percy Bengough a telegram accusing him of committing "treason" against Canadian workers.[12]

Charlie Murray said he knew of at least ten people who took payoffs to denounce former comrades in the labour movement. The most notable of these was Pat Sullivan, who, along with Murray, had led the 1939 fight to organize the fisheries workers in Lockeport. In March 1947 Sullivan appeared at a news conference in Ottawa and denounced his past activities in the CSU and the communist movement. Sullivan had held the second-highest labour position in the country, and his defection was a serious blow to the movement.

The press increased the public's fear of the communists. In March 1948 *The Halifax Herald* and *The Halifax Mail*—which Roscoe labelled as "just as foul a set of newspapers as is to be found in the world"—reprinted a series of twelve articles "exposing" communists in Canada.[13] The series, which had earlier appeared in *The Windsor Star*, gave its impression of the Communist Party: "Think of a group of cancer cells, multiplying incessantly, undeterred by the cutting away of any part of their mass, growing until they have corrupted and overcome the entire body in which they propagate, and you will have a pretty good idea of the revolutionary machine Karl Marx dreamed of and Lenin inflicted upon the world."[14]

Two months later, in the near-hysterical environment that was developing, the Halifax District Trades and Labor Council outlawed communists as members.[15] A.R. Mosher, president of the Canadian Congress of Labor, wrote in *The Halifax Herald* that Communists should not be entitled to Canadian citizenship and should instead have to be registered as agents of a foreign power. The attacks on suspected communists in Nova Scotia unions continued for several years.

Many people paid a high price in their business and personal life for their belief in radical politics. One of them was Jimmy Bell, one of the region's most dedicated labour leaders. Bell had played a prominent role in the formation of his union, the Marine Workers Federation, in Saint John and Halifax in 1945. For a short time following the war, he was active in the LPP. In 1952, after allegations that Bell, then secretary-treasurer of the Marine Workers, was a communist, the Nova Scotia Labor Relations Board ruled that Bell's shipyard workers union could not organize workers in Pictou, Lunenburg, and other places. Roscoe, furious over the Board's decision, wrote that only union leaders with the character and ability of people like Jimmy Bell organized new unions—while the labour establishment was only interested in raiding these unions after others had done the hard work. After Bell's union was barred the workers were organized by the United Steelworkers.[16]

Bell was removed from his prominent position in the labour movement and was reinstated only in 1965. He became secretary-treasurer of the Nova Scotia Federation of Labor, but he was still not considered "clean" enough to hold a membership in the New Democratic Party. He applied for and was denied membership in 1962. A letter from Allan O'Brien, NDP provincial secretary, informed Bell: "In view of your past association with a particular political party, and your continued association with that party in the common mind, your application has been rejected."[17]

In the Valley, Roscoe's arch-rival and the leader of anticommunist forces was Watson Kirkconnell, the president of Acadia University. At the height of the Red Scare in 1948, the university, with its strong Baptist tradition, appointed the virulent anticommunist as its president. Kirkconnell had come to the valley from McMaster University in Hamilton, where he was well known for his enthusiastic anticommunist activities. The Communist Party labelled Kirkconnell a fascist for his association with right-wing Eastern European groups, but he strongly denied the charge.

During his tenure as president of Acadia, Kirkconnell devoted many hours to sniffing out suspected communists and kept files on the activities of various people, such as Roscoe. He carried on correspondence with high-level officials in the RCMP, including the superintendent of the Special Branch, George McClellan, and former Communist Party infiltrator Jack Leopold, sending them names of people—including students—who may have been associated with communists. Kirkconnell also helped the RCMP recruit suitable members for its Special Services Branch.[18]

The McCarthy era efforts to expose Reds in both Canada and the United States was so wide sweeping that Roscoe thought he might be arrested. He warned members of his family and his friends not to be surprised if he was picked up by the RCMP. He had already lived through two Red Scare periods, and he believed that this one was by far the most dangerous. Roscoe was also concerned about the safety of Ruth and his two grandchildren in Detroit because she was speaking out about U.S. foreign policy. "Any of your neighbours would probably describe you as a Moscow agent. When you talk you are sure to go too far and they will not permit criticism. I insist that you must keep still."[19]

Despite his concerns for his own safety, Roscoe continued his critical writing. He accused Nova Scotia Premier Angus Macdonald and Attorney-General Lauchie Currie of Red-baiting to avoid being criticized for the increase in unemployment and prices in Nova Scotia in the late 1940s. Roscoe said they should be concerned about the state of the economy rather than spending their time worrying about communism.

> The people must not be allowed to think too much about the rapidly rising cost of bread, milk, butter, eggs, clothing, so these gentlemen shout incessantly, "the Bolsheviks will get you if you don't watch out." Their Labor Relations Board recognizes company unions, scab agencies, as bargaining agents for the workers, revoking the rights of bona fide trade unions, still singing the lullaby "the Bolsheviks will get you" if you belong to real labor unions.[20]

Roscoe believed that a major force behind the expulsion of so many communists from the labour movement was the CCF. He wrote that the CCF had "slid a long way into the slime" since the days of J.S. Woodsworth. "[M.J.] Caldwell and [David] Lewis are as sleek and oily and sly as latrine rats," charged Roscoe, "and have systematically expelled all socialists from the CCF. They have used their stooges in the unions to control or wreck those organizations."[21]

According to Roscoe's RCMP files, only twenty-eight people showed up for one of the few LPP meetings held in Halifax in 1948. By 1949 the impact of the Red Scare could be measured by the reception LPP party leader Tim Buck received during a visit to Centreville. Although it was mainly a social occasion, Buck gave one speech at the Centreville community hall—where ten years earlier he had packed the hall to overflowing. Now only about twenty people turned out, the majority of them members of the Fillmore and Sim family and their friends.

With the decline of the Labor-Progressive Party, Roscoe felt there was no other political party in Canada worth supporting. He particularly detested the CCF. In 1949, after CCF candidate Lloyd Shaw had been defeated in a by-election in Digby-Annapolis-Kings, Roscoe wrote Shaw a lengthy letter explaining that even though Roscoe and his family and friends had voted for him, Roscoe was opposed to CCF policies. Including the letter in his column in the *Steelworker and Miner*, Roscoe said that he would have felt much better about giving his vote to the CCF if the party had a cleaner record in foreign policy and Red-baiting. In terms of its foreign policy, he criticized the CCF for its general endorsement of the British Labour government policies, which he said were not calculated to foster peace but to build the power of American big business. He told Shaw that the CCF could not endorse the policies of Wall Street and still call itself socialist.[22]

❧

During the late 1940s Roscoe became increasingly obsessed with what he felt was the aggressive way in which U.S. military power and capitalism were being used to try to dominate the world. Often addressing the United States as "Uncle Shylock" instead of the usual Uncle Sam, he wrote more than two dozen angry columns in the *Steelworker and Miner* denouncing U.S. imperialist ambitions. In 1947 he wrote in the *Steelworker and Miner* that victory in World War II had cost the United States very little in terms of casualties and, with its industrial power and the A-bomb, America was now convinced it was invincible. He wrote that Wall Street, the real ruler of the United States, in talking about "Red Imperialism," was setting up a straw man to camouflage its own aggressive designs.[23]

Roscoe saw the Soviet Union as being in the vanguard of the nations challenging Wall Street's right to rule the world.[24] He thought that the fact that the United States was the only power with the bomb was a serious threat to world peace. In 1947 he wrote: "I believe that only when Russia has worked out the A-bomb for herself will there be a chance for a meeting and agreement in which the U.S.A., not having the whip hand, will talk and act like a gentleman, not like a gangster."[25]

Roscoe's protectiveness of the Soviet Union led him to write articles that indicated an unrealistic faith in that country. For instance, when the American Jewish Committee stated that there was widespread anti-Semitism in the Soviet Union, Roscoe wrote in the *Steelworker and Miner* that the charge must be a lie. "The Soviet Union is the only country in the world in which anti-Semitism is impossible," he wrote. "For thirty-four years, almost two generations, racial discrimination has been illegal in that country so that complete equality of all peoples has become second nature to the people."[26] It is unlikely that Roscoe was naive enough to believe this statement: He was probably again trying to protect the reputation of the Soviet Union.

Roscoe's greatest concern was not the state of affairs in the Soviet Union but the undercurrent of fascism in U.S. political ambitions. He came back to this theme time and time again in his column in the *Steelworker and Miner*, sometimes in a near-hysterical manner. Roscoe had seen how Germany had attacked Communists during the early 1930s, and he feared that the anticommunist campaign building in the United States was the first sign of a country on the road to fascism. He disliked the way the United States did business with the fascists who had been its enemies during the war. He was concerned about how it had begun to prop up undemocratic regimes in many parts of the world, including Italy, Greece, Turkey, Spain, Argentina, Iran, and China.

Roscoe believed that the McCarthy-led, anticommunist campaign was another element in the growing trend toward fascism in the United States.

> Behind the campaign is the most dangerous, cold-blooded and murderous gang of bankers, industrialists, politicians, militarists and church dignitaries ever to menace the peace of the world. They are far more dangerous than Hitler ever was because they are far more powerful. On their side is ranged every living fascist and Nazi and collaborator in every country. The United States is the heir to all this filth and is providing a worthy successor to Hitler and Mussolini.[27]

Roscoe also attacked racism in the United States. In 1947—long before the civil rights movement became popular—he used his column in the *Steelworker and Miner* to defend American blacks. He said that the United States had disenfranchised its twelve million blacks and treated them as "inferior animals." He said that Georgia and Mississippi were "cesspools" of racial hatred, and that until this was changed the United States had no right to pose as a democracy.[28] Roscoe was a great admirer of Martin Luther King Jr., and in later years exchanged letters with the U.S. civil rights leader, apparently telling him among other things that U.S. blacks should refuse military service until their economic conditions were improved by the government.[29]

Roscoe's reaction to the outbreak of war in Korea in 1950 was predictable and strong. He saw the conflict as a result of U.S. imperialism and wrote vicious articles attacking U.S. policy. Soon after Christmas 1950 Roscoe wrote: "While the rich churches of America celebrated the birth of Christ, whom they hypocritically call the Prince of Peace and pretend to worship, thousands of tons of jellied gasoline were showered on the Korean villages and people to burn them to a crisp. Now we are told that ten million of these people, who by the way are only 'God-damned Gooks' to the American forces, are on the roads as refugees."[30]

Roscoe's bitterness toward U.S. foreign policy made him a strong Canadian nationalist in the 1940s, long before it was a popular sentiment. But he wrote that the Canadian economy was so strongly controlled by U.S. big business that it would be almost impossible for Canada to regain its independence. Referring to the two countries as "United States and its poodle," he ridiculed succeeding Canadian prime ministers for "trolling to Washington to receive Wall Street's orders.... Canada has been taken over lock, stock, and barrel by Wall Street."[31]

❧

Sadly for Roscoe, the decline of radical socialism in Nova Scotia coincided closely with the death of three people who had been at the centre of his life: his closest socialist friend, his father, and his stepmother. Selina, who had been suffering from pernicious anaemia for years, died in Detroit in 1949 at the age of eighty-nine. After the funeral Roscoe and Margaret discovered that Willard wanted to return to Albert, where, as Roscoe said, he could be among preachers and praying people for the last years of his life. When they couldn't find a place in Albert for him, Willard decided he wanted to move to Centreville where he would live with Roscoe and Margaret. Even though Margaret was strongly opposed to the idea, Willard arrived in Centreville in the summer of 1951. Suffering from rheumatism, thrown into a strange environment, and not on particularly good terms with Margaret, Willard was not happy in Centreville. He died there the same fall, three days before his ninetieth birthday.

Roscoe's closest friend, Jimmy Sim, had become very ill during the winter of 1950. The Sim farm had not done well in recent years: He had worked the farm for nearly forty years but was close to poverty when he died in May 1951. The funeral was held in the living room of the Sim home, where—as Roscoe said in a tribute to Sim—they had discussed the world for thousands of hours. The local Baptist Church was asked to conduct the service, as a concession to the community and some members of the family. During one of their get-togethers many years earlier, Sim, Roscoe, and poet Ken Leslie had made a pact that when any one of them died the others would be there to bury them. Leslie, who returned to Nova Scotia from New York in 1949, sang a hymn at the funeral.

In his eulogy Roscoe said that Sim believed that the peoples of the world were so inextricably bound up together that an injustice to the people of Ireland, Greece, China, Indo-China, or Korea affected people even in Nova Scotia. He said Sim believed that people could not build a wall around themselves and exclude the changing world, and he touched on the impact of the Cold War on people such as Jim Sim. He said that communists, like Sim, were charged with everything in the calendar of crimes. "But can you seriously imagine our neighbor and friend, Jim Sim, a murderer, a torturer, a sinister schemer to enslave mankind? Or the dupe of such people? He had an enquiring mind and was not easily fooled."[32]

At the back of the room throughout Sim's funeral, unknown and unrecognized by the mourners, was a member of the RCMP who duly filed a detailed three-page report. He wrote that although the Baptist minister, Rev. G.W. Howard, did not approve of Roscoe speaking at the funeral, the Reverend felt he could not object since the service was in a home and not in the church. A second RCMP report on this matter, a month later, said that Roscoe's fifteen-minute talk was ostensibly a personal tribute, "but actually he was taking advantage of the opportunity to expound the Communist doctrine for the benefit of those present."[33]

When Roscoe stood before the small group of people during the funeral of his friend Jimmy Sim and said that Sim was no one's dupe, he knew in his heart that he was not telling the truth. While he had not spoken his mind publicly and had let only his closest friends, including Sim, know his true feelings, Roscoe had been aware for some time that Stalin's regime was brutal and totalitarian. It was a bitter pill for Roscoe who, since his emotional May Day in Petrograd in 1923, had cherished his vision of the Soviet Union as the salvation of the world.

By 1950 the attacks on communists had effectively destroyed the LPP in the Maritimes. The party that Roscoe had so proudly helped to organize at its founding convention in Toronto, and the party for which he had such great hopes, had nearly disappeared in the Maritimes. In view of the decline of the party, his dissatisfaction over how Ethel Meade had managed affairs, and his displeasure with the Soviet Union, he gave up his membership in the Labor-Progressive Party sometime around 1950. Roscoe was sixty-three years old. He had devoted some forty years of his life to establishing radical socialism in the Maritimes. Now that it was clear that socialism was not gaining the expected foothold, he was deeply saddened and disappointed. Roscoe's life had reached a turning point. While he still followed world events and sent off letters to the editor, he was alienated to the point that he never again became involved in party politics. Now, in his sixties, the success that Roscoe had longed for in his political life was to come to him in his other field of interest: horticulture.

SEVENTEEN

Mr. Green Thumbs

By the early 1950s Fillmore's Valley Nurseries had grown to become the largest nursery east of Montreal, selling more plants than all other Maritime nurseries combined. The rapid expansion of the business was largely the work of Roscoe Fillmore's son Frank, who brought an aggressive style of management to the company. Although Frank had no formal education in horticulture, he had learned the fundamentals of gardening from his father and his brother Dick.

The nursery at Centreville had become more of a family affair than ever before. Margaret, Rosa, Irene, and Irene's mother Cora Cunningham, along with other relatives, joined the regular staff during the spring and early summer to package plants for shipment throughout the Maritimes. In the summer they worked in the fields, down on their knees, pulling weeds. Margaret, now in her fifties, could be seen dressed in overalls, rubber boots, and a wide-rim hat, down on the ground, transplanting or weeding. She enjoyed the work. On Mother's Day, one of the busiest times at the nursery, Margaret could look elegant, dressed up in her best clothes, with bright bows, long strings of colourful beads, a new straw hat, and her beautiful dark eyes and smile flashing at the customers.

Roscoe and Margaret's house was located right in front of the main nursery buildings. Purple and white lilacs, beds of bright flowers, and a few of Roscoe's prized rhododendrons adorned the property. A large shed, where all the packaging was done, and two large greenhouses were behind the house. Beyond the greenhouses there were fields of plants stretching for a half-mile to a forested area. Located next to the nursery was a side attraction for visitors: Charlie and Mabel MacDonald's concrete house, which was surrounded by more than a dozen life-like statues, including deer and a moose.

Roscoe, who celebrated his sixty-fifth birthday in 1952, devoted his attention to plant propagation, which he preferred to the frustrating work of helping to manage the nursery. He supervised the growing of new plants—from seeds, cuttings, and grafts—that were needed to keep the nursery supplied. But beyond this Roscoe conducted experimental work to develop new strains of plants that would survive the cold winter and short growing season of the Maritimes.

Roscoe had set up his own little propagation lab in the main greenhouse, and he spent hour after hour refining ways of developing new strains of plants. When his friends visited they would often join Roscoe at his work-bench, talking about the latest political developments while Roscoe carried on with his work. He showed remarkable creativity and patience in the slow, painstaking propagation experiments. To begin his work, Roscoe gathered plant specimens locally and from Ontario, Holland, and, later, from a nursery in Iowa where Dick worked in the early 1950s as a plant propagator. He tried to keep up on the latest propagation developments in the United States and obtained books and pamphlets on Soviet work through the Soviet Union's information office in Washington. Using various methods arrived at through trial and error, Roscoe began to develop new strains of plants, including holly, magnolias, hydrangea, Colorado blue spruce, rhododendrons, and azaleas.

Of all the plants Roscoe worked with, rhododendrons were his favourite. He had begun experimentation to develop hardy new rhododendrons in the early 1940s, a time when the plants were not available commercially in the Maritimes. There were probably only three plantings of the shrub in Nova Scotia, one in Yarmouth, one hidden away in a swamp near Annapolis Royal, and one in the Public Gardens in Halifax.[1] Roscoe wanted to develop a plant hardy enough to be sold by the nursery to people living in most parts of the region, and he used several different methods to do this. One method was to import dozens of plants from Holland, where the climate was much milder than Nova Scotia, and plant them unprotected in the fields.[2] Each winter, year after year, Roscoe watched carefully as nature weeded out the least hardy. Each winter at least one or two plants hardier than the rest would survive. He kept the seeds from these survivors, planted them, and repeated the process over and over until plants that were sure to survive had been developed.[3]

Roscoe tried other methods that involved more creativity. Acting on a suggestion from Dick, he began to try to propagate hardy rhododendrons from single leaves cut from a healthy plant. This method involved creating a perfect environment in which the leaf, put in what might be a mixture of sand and soil or vermiculite, would be able to develop its own root system. To provide a constant level of heat, a special heating cable was placed in the greenhouse. As many as a thousand cuttings were planted at one time. To his delight, Roscoe discovered that several varieties would root readily. The second year, many of the tiny plants were killed by a fungus, so he experimented with different ways of protecting them, finally discovering that a thin layer of polyethylene thrown like a tent over a bed of the plants provided the best growing environment.[4]

He used yet another method on a particular variety of rhododendron he had seen in Boston during the war. The Carolinianum was a small rhododendron, with a small pink flower. Roscoe thought they were exceptionally beautiful and brought some of the plants home to Nova Scotia. Not unexpectedly, they were killed by the harsh winter weather. Then he bought seeds of the same species, collected in the wild in the Carolinas. He germinated about one hundred plants, keeping them in the heated greenhouse for the first winter. In the second winter he kept them in a special cold frame, which partially protected them from the winter weather. His idea was to expose the rhododendrons gradually to more and more cold and hope that by the end of the process some would survive. Ken Wilson, brought from England to help with the nursery's growing volume of propagation work, said that when he arrived in Centreville in the spring of 1954 there were about twenty surviving plants, each about a foot high. By this time the plants had been moved outdoors and were in a lathe house, which provided only slight protection from the winter. But then, Wilson recalled, Nova Scotia had a very harsh winter and only seven of the plants survived. According to Wilson, this was just what Roscoe wanted. Now any seedlings from those final survivors would be just as hardy as their parents. Roscoe put most of those plants in an open field, where they flourished.[5]

Roscoe devoted thousands of hours to the work of nurturing the plants, losing most of them to both heat and cold, until he had developed a broad range of varieties and colours that would survive. Throughout the propagation process, Frank helped by designing new equipment, such as plant trays and watering systems, that was suited to the particular plants. Often, as a final test to determine the hardiness of his new plants, Roscoe would send specimens to friends throughout the Maritimes and have them report back to him a year or two later.[6]

Over a period of twelve years, Roscoe identified and helped to develop about forty rhododendron hybrids that were hardy in Nova Scotia. From these he and Frank selected about twenty for sale at the nursery. Roscoe did similar work with azaleas. Largely because of this work it became possible by the mid-1950s for people in many parts of the Maritimes to grow both rhododendrons and azaleas.

Roscoe, always one to look at the larger picture, thought that this kind of research work should have been done by the Agricultural Research Station at Kentville. He was critical of the research station for not getting involved in growing ornamental materials, and in the mid-1950s he finally succeeded in getting the farm interested.[7] The station purchased six of his large rhododendrons and six azaleas, along with others of his plants, and Roscoe helped Donald Craig, assistant director and head of the crops section at the farm, to establish the first plantings at the Kentville site. The large collection of rhododendrons and azaleas, many of them developed by Craig, became a major attraction for visitors.

❧

To attract more customers, the Fillmores began promoting the nursery on the Kentville and Halifax radio stations. But instead of presenting long lists of plants and their prices, they aired answers to common gardening questions, such as lawn care and soil preparation. The number of visitors to the nursery continued to increase.

The success of the radio spots, and the fact that the nursery was flooded with letters from gardeners who had questions, made Roscoe start to think about the value of producing a gardening book that would interest ordinary people in gardening, that would appeal to homeowners who didn't have a great deal of money to spend on home beautification. He knew that while the wealthy hired the services of landscape architects, most people had to depend on their own imagination and information gathered from books, garden papers, catalogues, and nursery workers. Looking around, he saw an absence of adequate Canadian gardening books. For instance, the only books on roses—which came from Britain, New Zealand, and the United States—tended to ignore Canada. Roscoe believed that people needed the opportunity to acquire the necessary knowledge about their own particular climate, and how plants could grow in it. He would avoid technical language and, wherever possible, use common names for all plants.

His theory was that for a great many people the garden provided a much-needed escape from the mechanical existence of modern life. He lamented the way urban development had taken people away from the countryside. In his book *Roses for Canadian Gardens*, which would come later, he wrote: "In a matter of sixty or seventy years the towns and cities have taken over to such an extent that less than fifteen per cent of us have to do with cultivating the land. Many of our sons who were wild for city living now wish themselves back on the land. This being impractical, the garden will have to make due as a substitute, a sort of escape valve. I suspect the garden is preferable to the tavern."[8]

When Roscoe wrote to Macmillan of Canada about his work he got a lukewarm response. Macmillan—a branch-plant publishing house—said they were interested in seeing the manuscript but warned that this kind of book couldn't be published in Canada alone because of the small national market; they would need their New York office to publish it as a North American book.[9] The Ryerson Press—a Canadian-owned publishing house that had been in operation for more than 120 years—was less tentative. Ryerson was "very much interested," wrote one of the company's assistant editors. "There is a feast of gardening books on the market but so many of them are either English or American and the conditions do not always apply to our Canadian situations."[10] Roscoe had found the publisher he wanted.

Green Thumbs: The Canadian Gardening Book was published by Ryerson in 1953 and provided a personal, general overview of all aspects of gardening. An excerpt appeared in *Canadian Homes and Gardens*, the country's largest gardening magazine. The book proved to be an immediate success. The CBC's "Prairie Gardener," Dr. A.R. Brown, said it was the most thorough book on gardening he had seen. In another review, *Canadian Homes and Gardens* praised the book for its "easy, almost conversational" writing style.[11] There were interviews on radio, appearances on television, and invitations to visit horticulturalists in different parts of the country.

Whenever he spoke, Roscoe was promoted as "Mr. Green Thumbs." Those attending his personal appearances would have seen a rather unconventional-looking older man. Never especially concerned about his appearance, Roscoe often wore an inexpensive, pinstripe jacket and vest with a dark shirt, a bow tie, and rumpled trousers. He was warm and friendly, always ready for a chuckle of some sort. The outward sign of the rebel that still lurked within, relatives said, was the unkempt silver hair that often stuck out from the back of his neck and curled onto his jacket. They also said he was ever mindful of returning to the poverty of the Depression. He took to carrying substantial sums of cash—sometimes as much as $200 or $300—with him, as if to reassure himself that the difficult days of the 1930s were gone forever.

About the time *Green Thumbs* was published, Roscoe retired from his job as gardener for the Dominion Atlantic Railways. He got a decent DAR pension, and his thirteen years of service entitled Margaret and him to a free railway pass for anywhere in North America. Roscoe took at least one major trip a year, often to see his relatives in Detroit and Tim Buck and other friends in Toronto.

The summer the book was published Roscoe went to Ontario to visit several nurseries and see what new gardening methods were being developed. One of his visits was to Woodland Nurseries, located in Cooksville on the outskirts of Toronto and owned by Leslie Hancock. Like Roscoe, Hancock was greatly interested in plant propagation. He had done research on rhododendrons and azaleas and had brought new gardening techniques to Canada from China, where he had taught biology at the University of Nanking in 1923. Perhaps even more important, Hancock was also a socialist, a kindred spirit. He had been a CCF member of the Ontario legislature until he and another member were expelled in 1946 for advocating that the CCF co-operate with the Labor-Progressive Party and the Liberals to attempt to defeat the Conservatives.[12]

Roscoe and Hancock, with both gardening and radical politics in common, got on famously. They continued to visit each other over the years. Roscoe provided Hancock with rhododendrons for propagation, and the Valley Nurseries bought plants from Woodland Nurseries. They both joined the International Plant Propagators Society, a network of plant propagators, and were the first Canadians to deliver keynote speeches at a Propagators Society convention.

Roscoe's oldest son Dick was one of the founders of the International Plant Propagators Society and one of the most respected plant propagators in North America. He had obtained a Master's degree in ornamental horticulture and plant breeding from Cornell University and had worked at Harvard University's Arnold Arboretum, which many horticulturalists thought was the best plant propagation centre in the world. After an unhappy experience at a commercial nursery in Iowa, Dick became chief horticulturalist at the Sarah P. Duke Gardens at Duke University in Durham, North Carolina, in 1956. His job was the envy of many horticulturalists. Dick played a leading role in transforming the gardens into one of the most admired ornamental gardens in the United States. The gardens at Duke covered fifty acres of landscaped and undeveloped area and included more than one thousand species from grass to towering pine trees. They were grand and luxurious compared to the small gardens at Grand Pre where Dick had worked before the war.

Much different than his fiery younger brother Frank, Dick shunned the world of commercial gardening. He believed that gardens provided basic emotional and spiritual needs. Dick, a published poet and a painter, thought that people needed the gardener as much as they needed writers and artists, "for beauty is one of the necessities of life."[13] Over the years there was a constant stream of advice, cuttings, and plants flowing between Durham and Centreville.

❧

Roscoe and Margaret had always wanted to tour the west coast. In early 1954 they left Nova Scotia by train for a trip to California and British Columbia. They saw the ancient redwoods of California and all the tourist sites of San Francisco. In San Jose, Roscoe spoke to the Canadian Club about gardening. Returning by way of British Columbia, they visited many of the province's beautiful gardens. But Margaret was not well during much of the trip. Stopping over in Detroit on their way to California, Margaret told Ruth that she was feeling ill, but she wouldn't say anything to Roscoe because she didn't want to ruin his trip.[14] Roscoe finally realized that Margaret was sick around the time they were leaving California, and they stayed several extra days in a hotel in Seattle so she could rest. It soon was apparent that she was very ill. For Roscoe the journey home seemed like the recurrence of a terrible nightmare. He relived a terrifying train trip from Maine to Saint John many years before, when, as a six-year-old boy, he had seen his mother become sick and return home to die in the Maritimes. Now he feared the same thing was happening to his wife.

After they got home Margaret was admitted to hospital in Kentville, where the doctors discovered she was suffering from cancer of the pancreas. The disease had taken hold and it was clear she would not survive. Within a few weeks, weak and tired, she stopped talking. Finally Margaret was brought home to Centreville, where she was cared for by a nurse until she died in her sleep, at age sixty, on August 2, 1954. Roscoe spent the last two days and nights with her in her room.

When the funeral was over Roscoe was full of guilt. He blamed himself for being away from Margaret too frequently during her illness. He missed her terribly. They had been married forty-four years. "Everything I see reminds me of her," he wrote to his daughters Allie and Ruth. "I suppose there were many days when we talked less than five minutes altogether but I am so conscious of her absence tonight that I scarcely know what I am to do."

Roscoe was grateful that at least he had told Margaret during her illness how important she had been to him.[15] During recent years Margaret and

Roscoe had experienced far fewer disagreements than during the tough years of the 1920s and 1930s. Roscoe's temper seemed less noticeable as he became older and his responsibility for supporting a large family eased. For many years, though, they had not expressed their feelings to each other. "Mother was not a very demonstrative person," Roscoe wrote to Allie and Ruth, "and I lived for years at a time wondering if she cared at all for me, and that was bound to affect my actions and attitude toward her. And then one day not so long ago she gave me proof of her love when I found her beginning to cry and went to the bed and took her in my arms and she kissed my hand. She said very little but I told her what she meant to me, and I took this as evidence that all our antagonisms and troubles of years ago were forgotten and dissolved. I know that the past few years she has been happy and content."[16]

❧

To fight off depression and his feeling of loneliness after Margaret's death, Roscoe moved in with Rosa and Gerry, who were living in a huge old farm house just down the road from the nursery. He also threw himself into his work. In the fall he went to Halifax, Saint John, and Moncton to give gardening talks on television with the idea of generating new business for the nursery. Ads were run in all the papers promoting "Mr. Green Thumbs." Visiting Moncton, Roscoe thought about how dramatically his life had changed from the early days. While forty-five years earlier in Moncton he had stood on street corners preaching revolution, this time he spoke to the conservative, community-minded Rotary Club. Roscoe was nervous, but his talk was well received. "I think they would have listened for two hours," he wrote to Dick.[17] He returned to Centreville with orders for $850 and the prospect of considerably more business.

As the nursery grew father and son frequently argued about how quickly the business should expand. Roscoe, who still had his generation's nagging fear of returning someday to the poverty of the Great Depression, criticized Frank for spending too much money. When Frank wanted to buy a machine that he felt would speed up the planting of shrubs in the field, Roscoe was against the purchase. Frank complained in a letter to Dick that Roscoe was no help in planning for the future of the nursery. "He vetoes anything that even looks like expense, regardless of any return. He fumed like crazy about the planter until I just dropped the matter and bought it on my own. Now he's like a kid with a new toy."[18]

Roscoe also criticized Frank for playing hockey. Roscoe, who in his own youth had cared very little about sports, couldn't understand why a grown man would endanger himself by engaging in an activity as violent as hockey. His greatest fears were realized on an evening in January 1955 when Frank, playing for the Kentville Wildcats in a game in Wolfville, was struck in the face by the puck. He lost the sight in his left eye. After spending three weeks in hospital and recovering at home, Frank resumed management of the nursery in the spring, though Roscoe felt that the injury interfered with the amount of work his son was able to do.

The good news for Roscoe was that *Green Thumbs* was such a success that it was going into a second printing. *Green Thumbs* outsold all other gardening books in the country, leading Roscoe to write a second book for Ryerson Press. *The Growing Question*, a question and answer book again aimed at ordinary gardeners, was published in 1957. The reviewer for *Canadian Geographical Journal* wrote, "This is a book that should be on the shelves of all Canadian gardeners."[19]

Over the next four years Roscoe published two additional books, *Roses for Canadian Gardens* (1959) and *The Perennial Border and Rock Garden* (1961), both of them well received by reviewers and the gardening public. Now his four books were promoted nationwide by Ryerson Press as the most complete set of books on Canadian gardening available, and they were displayed in Europe as part of a book fair sponsored by the federal government. While the books sold relatively well for the times, the royalties did not contribute significantly to Roscoe's income. He gave away so many free copies to friends and relatives that people joked that he must have taken a loss on the books. Roscoe partly completed a manuscript for a fifth book on fruit trees and fruit plants, but it was never published.

Partly due to the recognition he gained through the publication of his books, Roscoe was soon regarded as one of the top authorities on gardening in eastern Canada, frequently consulted by other gardening experts and local garden clubs and asked to write articles on his work. Much to his amusement, Roscoe discovered that his growing status as a leading horticulturalist had won him new respect among those members of his family who had for so many years frowned on his radical politics. He had, he said, been transformed from the black sheep of the family to a white-haired boy.

Plants had replaced politics as Roscoe's main interest in life. He loved to collect plants, and he would beg, borrow, buy, or steal them. Ken Wilson, who came from Britain to work with the Fillmores as a plant propagator, said that most of the unusual plants that Roscoe picked up would wind up in a lathe house in an out-of-the-way corner. Wilson remembered one time when a woman customer came into the packing shed with one of Roscoe's treasures, which she had dug up. She got a stern lecture. Don Craig of the

Experimental Farm recalled that if Roscoe saw a particularly lovely flower he would say, "Isn't that the prettiest God damn thing you have ever seen?"[20]

Beyond their work at the nursery, Roscoe and Frank took an interest in the life of Centreville. The nursery broke the colour barrier that had existed for many years in Centreville by hiring black workers from the nearby village of Gibson's Woods. Frank broke another colour barrier in Kentville by taking one of the black workers from the nursery to his regular barber shop for a hair cut. On another occasion, when Roscoe and Frank feared that a village bylaw was going to be passed to bar Jehovah's Witnesses from using the local community hall, they went to a village meeting and proposed that there should be no discrimination in the use of the facilities. When a vote was called, the result was thirteen to twelve against Roscoe's motion, with fourteen abstentions. Roscoe sent Dick an account of the evening: "And yet everyone of the damned hypocrites claimed they had no intention of discriminating!"[21]

❧

By the mid-1950s, Roscoe or Frank could drive on the main streets of most towns in the Maritimes and see the results of their work. Many of the trees, shrubs, and other plants that were in the parks and in people's yards came from the Valley Nurseries, and they had landscaped thousands of homes. Ironically, some of the most outstanding landscaping jobs the Fillmores did were for homes in the South End of Halifax, which were owned by the same wealthy families that Roscoe had criticized and condemned through his political activities.

As manager of the nursery, Frank did business with the Liberal administration of Angus Macdonald—the very government that Roscoe had heaped ridicule upon after the war. The Liberals awarded the nursery a contract to landscape the grounds of Keltic Lodge, the government-owned resort carved out of rock in Cape Breton. While Frank's personal politics were considerably left of the mainstream parties, on at least one occasion it was alleged that he urged employees of the nursery to vote for Conservative George Nowlan.[22] This occurred during the 1957 federal election campaign and was probably due to the fact that one of Nowlan's chief campaign organizers was Victor Thorpe, a Kentville lawyer who had been born and raised in Centreville and was the nursery's lawyer. For his part, Roscoe had no interest in supporting either the Tories or Liberals. While he still held radical political views—which were being expressed frequently at the time in the letters column of Halifax newspapers—Roscoe seemed to have no objections about doing business with most anyone who wanted to buy plants and trees.

Even though Roscoe was no longer active in the Labor-Progressive Party, the RCMP continued to put Roscoe and the nursery under surveillance. A typical RCMP report, prepared by Cpl. W.A. Taylor in June 1954, noted: "Patrols in the vicinity of [Roscoe's] home have revealed large numbers of cars there during the day and early evening. On May 24th, while in the district, it was estimated that between thirty and forty cars were stopped there selecting shrubs, flowers, etc. In view of this, and the fact that the customers come from all parts of Nova Scotia, it is difficult to separate visitors to his home from persons buying his goods, as they use a common parking lot."[23]

In fall 1955, when Roscoe drove to Halifax to speak with potential customers about gardening, the RCMP had him watched by two officers. The police kept a watch on the Lord Nelson Hotel, where Roscoe stayed, but reported that they saw no Labor-Progressive Party people visiting Roscoe. The police also did spot checks on Roscoe's outgoing phone calls and found that all of them were to people concerned with gardening.

The RCMP's surveillance of Roscoe illustrated the police's failure to provide criteria for determining who was a legitimate security threat and who was not. In the absence of a policy the force followed and filed reports on thousands of Canadians who presented absolutely no threat to the country's security. Early on, when he was a member of the Prime Minister's Office during World War II, Jack Pickersgill analysed an issue of the RCMP's *Intelligence Bulletin* and noted an "anti-Red complex," no discrimination between legitimate political criticism and subversive doctrine, a distinct inability to distinguish between "facts" and "hearsay," and no evidence of any suspected sabotage or espionage.[24] In dealing with Roscoe, the RCMP did not distinguish between his right to express his views—no matter how radical they were and no matter how much the RCMP might disagree with them—and his involvement in activities that were a threat to the country.

❧

If the RCMP had been monitoring Roscoe closely it would have learned that he had become disturbed by many things he saw happening in the Communist Party and in the Soviet Union. Earlier, Roscoe and many of his friends—perhaps embarrassed about having been deceived by the party and Stalin—had not wanted to talk about the past. But in the 1950s Roscoe broke his long silence by criticizing the Soviet Union to relatives, friends, and neighbours. In 1956, following the invasion of Hungary by Russian troops, he showed his true feelings about Stalin and the Soviet Union in a letter to his daughter-in-law, Becky Fillmore: "If the Hungarian affair means the Soviet Union is again using the iron fist, I will oppose her and criticize [her] freely and bitterly."[25]

Roscoe was one of many hundreds of independent communists who were angry over the Canadian party's continued adulation of Stalin. After Nikita Khrushchev delivered his 1956 secret speech about the brutality that the Soviet people had faced under Stalin, Roscoe was critical of the Canadian party—which was seriously damaged over the issue—for its lack of action on what had taken place in the Soviet Union. Recalling his disappointment, he wrote in his autobiography in 1959: "Those who are sincere in their pleas that they were terribly shocked by Khrushchev's speech have been the victims of a mechanical acceptance of faith of dogmas, interpretations and tactics that were really a recognition of the infallibility of Joseph Stalin, and this was no whit more intelligent than acceptance of the Doctrine of Infallibility of the Pope."

It appears that it may not have been until 1964 that Roscoe criticized the Soviet Union publicly, accusing Soviet officials of being too sensitive to criticism in a letter to *The Canadian Tribune*. That same year, after Khrushchev was replaced as party leader, Roscoe wrote a terse letter to the executive of the Canadian Communist Party complaining about the way Khrushchev had been treated. Later, in 1968, less than three months before Roscoe's death, the Soviet Union sent troops into Czechoslovakia to crush the Prague Spring. Disappointed again by the Soviet Union, Roscoe wrote his granddaughter Barbara, "It's amazing how quickly people can become conservative after a revolution."[26] Roscoe protested the action by writing the Soviet embassy and the Communist Party. He also sent a short letter to *The Canadian Tribune* and seemed surprised that it was published.

> At 81 years of age, and after over 50 years of defending in general the action of the Soviet Union, I find myself unable to defend her actions and policies in the occupation of Czechoslovakia. I cannot defend this action for I believe it will drive millions of people into reaction and into hatred against the Soviet Union and the communist parties throughout the world.... Evidently a course of liberalization is badly needed in the Soviet Union.[27]

Roscoe's disappointment with the Soviet Union led him to turn to Cuba and China as countries that he felt were providing a better communist role-model for the world. He planned a visit to Cuba for 1960. He also spoke frequently during the 1960s of moving to China and spending his last years there.

While Roscoe had no doubt that he was right to speak out against the

Soviet Union, he still had a nagging sense of guilt about attacking what he saw as the wellspring of the communist movement. In defence of his criticisms he wrote in his autobiography:

> I believe that I am no less a communist today than five, twenty or forty years ago just because I cannot accept a discipline that to me lacks intelligence. I shall always be a communist but I reserve to myself a certain autonomy, the right to be critical of those tactics and policies with which I cannot possibly agree. If the Communist Party in Canada were outlawed again [the LPP changed its name back to the Communist Party in 1959], I shall hasten to rejoin it, if permitted.

The collapse of his two greatest wishes for life—communism for the Maritimes and a free, more liberal society in Russia—did not leave Roscoe cynical. He maintained a great interest in all political affairs and, for the most part, was optimistic about the future of humankind. During one contemplative mood he reflected on the theories of Nazi biologist August Weismann: "Maybe Weismann was correct—man can learn to make all the gadgets including the bomb to abolish himself—this is his progress, but he cannot learn to treat his fellow man with decency. Perhaps it's just not in his genes and chromosomes. If I believed that I would conclude that I had wasted some 55 years of my life."[28]

In his unpublished autobiography, Roscoe reaffirmed his belief that some day there would be a co-operative society. He wrote that a competitive society was responsible for most of the world's troubles. He believed that once civilization had tired of poverty and recurring wars, co-operative society would be introduced.

EIGHTEEN

The Man Within

WHEN HE WAS in his sixties and at the height of his gardening career, a nostalgic force began to pull Roscoe Fillmore back to his childhood home in Albert County, New Brunswick. It had been more than fifty years since his family had been driven from the backwoods village of Lumsden by a deadly typhoid epidemic. Roscoe, now thinking more and more about his past, was curious to see what remained of his birthplace.

On a Labour Day weekend in the early 1950s, Roscoe and other members of the family got in the car and headed for New Brunswick. After a brief stop in the village of Albert, they drove along what remained of a dirt road that led into the woods toward Lumsden. The hills were glorious with their early autumn colours. The road climbed up along a gorge, and down at the bottom they could see Crooked Creek. Roscoe recognized pools in the river where he had fished as a boy. As the car began its winding climb up the side of Lumsden Mountain, Roscoe found himself talking in hushed tones, not sure what he would find—or what he wanted to find. As he drove, he fought off his emotions, thinking of the last time he had been on this road—as a small boy, riding on a wagon carrying the body of his mother, who was to be buried in Albert.

Roscoe knew that lumbering crews had used the abandoned houses during cutting season and that many of the homes had been destroyed by fires. Now, when the family reached the plateau where Lumsden once was, they got out of the car. They could see only forest growing around them. A flat area that Roscoe thought he remembered as being a field of oats now nurtured trees that had reached forty feet. They started to look for the signs of old buildings, but nothing was left standing.

Despite the drastic changes, Roscoe found that he had imprinted in his memory a map of the small village as it was in his early childhood. He wrote in his autobiography: "I was bound for the place where my Grandfather John Fillmore had lived, for this had seemed the centre of my small world when I was young.... I had looked upon Grandpa John's place as home, and I knew I would be oriented and know my way around once I found where he had lived. I went to it unerringly, though I believe I had not been there since my seventh year."

All that was left of his grandparents' house was the bare outline of the foundation. Roscoe's mind raced back to when he was a small boy and Grandpa John had taken him on his knee and given him peppermints. He vividly recalled the delicious jelly Grandmother Elizabeth made from the high bush-cranberries that grew beside the back door of the house and the time that he and his sister Ellida had followed a surveyor's snowshoe tracks into the forest where they were later found by their near-frantic mother.

They got back in the car and drove further along, watching for the Keillor Road. After some searching, Roscoe found what he thought was once the driveway to his family's house. He walked in from the road and explored the thickets where he thought the house would have been. Nothing. Eventually he sat down and wondered why he couldn't find the foundation. The answer came to him suddenly. He wrote later: "We all remember going back to places we knew in our childhood. To children's eyes, they had been imposing. To a six-year-old a hundred yards is a long journey. We go back to see it sixty years after and it is a very humble thing."

Roscoe began to rethink his search. He realized that no settler would have built as far from the road as he was looking. He remembered the hard winters and the problem of keeping the roads and lanes clear of snow. He moved back toward the road, squeezing through the fir thickets, and found what he'd been seeking. At his feet he saw the foundations outlined by a low dyke-like bank. Spruce seed had germinated on what had been the cellar floor and twenty-foot trees now stood where once had been the little house where Roscoe had spent his first few years.

Roscoe, fighting to hold back the tears, began to romanticize about "the good old days" in Lumsden, but he quickly realized this was a mistake. He saw how the forest was closing in on the few remaining signs of the village. He thought it was right that nature worked to cover all scars.

His visit to Lumsden instilled in Roscoe a great interest in his family's history, and he spent hours in libraries and archives in the Maritimes researching his roots. Roscoe's cousin, Herman Fillmore, who also was born in Lumsden and had lost his father in the typhoid epidemic of 1894, was even more deeply involved in researching the family's background. In a cemetery in Jolicure, a tiny hamlet near Aulac, New Brunswick, where the ferry connects with Prince Edward Island, Roscoe and Herman found the gravestone of their great-great-grandfather, born in 1764. Further work uncovered the fact that their ancestors had come to the Maritimes from Connecticut, most likely in 1763 as part of the wave of New England Planters who had occupied the marshlands along the Bay of Fundy.[1] Roscoe took great pride in the fact that he was a fifth-generation Fillmore born in the Maritimes.

❧

Roscoe's last years may have been the most content of his life. Family members began to see a distinct change in him. He tended to keep his temper in check and be cheerful and kind to everyone in the family. Rosa believed he was trying to make up for the things he had not done in the past. The improved disposition may have been in part due to Rosa's devotion to him since Margaret had died. All four walls of his small bedroom were stacked to the ceiling with books. He would read well into the night and sleep until noon the next day, when Rosa would serve him grapefruit, toast, and tea. He also liked old cheese and, when it was available, caviar.

In summer, wearing a wide-brim hat to protect him from the sun, Roscoe would sit in a rocking chair on the lawn and read for hours, with dozens of magazines and books strewn around him.In the evening he would sit in Rosa's living-room, watching television, pausing occasionally to read a book he had on his lap. He cut down on cigarettes and took to smoking a pipe. Sometimes he would have a drink of rum and Coke. He would listen to records. Roscoe's taste in music was influenced by his politics. He was particularly fond of black American singer Paul Robeson—he had been furious when Robeson was barred from entering Canada in 1956. Roscoe also liked to listen to recordings of the famous speeches of prominent activists, especially a recording that Martin Luther King Jr. had made at the Canada-U.S. Peace Bridge.

Roscoe continued to travel. He took his most exciting trip in 1958 when he visited Europe for the first time since 1923, when he had gone to Russia. Roscoe and his daughter Allie and her family travelled through West Germany, Switzerland, France, and England, and Roscoe sent home accounts of his European gardening discoveries and of an emotional visit to Karl Marx's burial site in London.

Roscoe got more enjoyment out of his grandchildren than he had from his children when they were young. He attributed this to the fact that when he was younger he was so preoccupied with trying to earn enough money to keep food on the table, and with his political interests, that he had little time for his children.[2] Now he went to great lengths to either write or speak to most of his eight grandchildren about politics and the development of society. He saw this as an attempt to preserve his outlook and pass it to a third generation. "These young people are my immortality, my excuse and justification for living," he wrote.[3]

Dick's daughter Mary, a precocious little girl who lived with her parents in Durham, North Carolina, was especially receptive to Roscoe's political views. Roscoe wrote her frequently, and whenever she visited Nova Scotia the two of them would talk long into the night after everyone else had gone to bed. Later in life Mary—as a university student and as a feminist who directed affirmative action programs—continued to follow roughly the same kind of political path, adding her own analysis. "I do feel very much that he set me on a certain track," Mary Fillmore said, "and it was a critical track. It was a track that questioned whatever I read or saw, and a track that caused me to be as skeptical as possible for as long as possible about what was being said to me as an interested citizen."[4]

Friendships were important to Roscoe. He spent time with many old friends from his activist days, but his favourites may have been Kaye and Charlie Murray, who returned to Halifax in 1956 from New Brunswick. Roscoe would often turn up at his friends' places with a gift of one of his favourite plants—perhaps an azalea or a flowering shrub—or a copy of one of his books with a personal dedication. In the early 1950s he had renewed his friendship with poet and magazine publisher Ken Leslie, who had returned to Nova Scotia from the United States. Leslie was not a Red, but he had been a prominent figure among American liberals and had come under FBI surveillance for possible "un-American" activities. A power struggle at *The Protestant*, the magazine that Leslie had led since before the war, and turmoil in Leslie's personal life contributed to his decision to return to Canada. His return must have been a demoralizing experience. In New York he had been a highly regarded crusading editor and political activist. In Nova Scotia Leslie continued to publish a considerably diminished version of *The Protestant* and a succession of other small magazines aimed primarily at the United States, and he kept on writing poetry, but he was not widely known or recognized for his work.

Roscoe was pleased that Leslie was back home. Whenever the two men got together they had heated but amicable arguments over politics and religion. Leslie's strong religious faith was often at the centre of their disagreements. Leslie saw no contradiction in his political and religious beliefs, whereas Roscoe remained a lifelong opponent of any kind of religious doctrine. Despite their arguments, Leslie was a keen supporter of Roscoe's project to write a book on gardening, and he read the manuscript of *Green Thumbs* and offered comments. The time that Leslie spent with Roscoe and his other socialist and communist friends appears to have influenced his writing, which was more political and more angry than before. In total, Leslie published six books of poetry. Despite acclaim abroad, he was deeply disappointed that his work received little attention in Canada. When he died in 1974, he had been living in a nursing home. Only about a dozen people showed up at his funeral.[5]

❧

Enjoying his new celebrity status as a gardening authority, Roscoe was in a cheerful mood when left Nova Scotia by train in November 1956 to give a talk at the annual convention of the International Plant Propagators Convention in Cleveland. Before going to Cleveland, he was looking forward to a visit with relatives in Detroit. But when he arrived at the Immigration Department booth in Windsor, the officer on duty seemed to take a long time shuffling papers and consulting someone else. Roscoe grew impatient. After what seemed like several minutes he was taken aside and another official started asking him questions. Strongly believing that people shouldn't be interrogated at border crossings, Roscoe bristled. When the officer asked, "Are you, or have you ever been, a member of the Communist Party?" Roscoe shot back, "That's none of your damn business!"

Roscoe was sixty-nine and looked more like an elderly man on his way to see his grandchildren than like a treacherous radical. He was so indignant about being asked about his past that he refused to answer any questions. He was politely informed that he was denied entry into the United States. It was still the McCarthy era, and the Americans had passed a law that banned foreigners from entering the country if they had unacceptable political beliefs. Though he was no longer active in the Communist Party, Roscoe realized that the RCMP had apparently given information from his file to the FBI.

Angry and outraged, Roscoe was escorted back to the Canadian side of the border. After seeing Ruth and his two grandchildren, who came to Windsor to console him, Roscoe arranged to take the train back to Nova Scotia. On the ride home he was depressed over his treatment at the hands of U.S. officials and perhaps feeling a little sorry that he hadn't answered their questions.

Once he was home, he asked George Nowlan, Conservative MP for

Digby-Annapolis-Kings, to see if he could persuade Liberal justice minister Stuart Garson to intercede on his behalf with U.S. immigration officials. Nowlan contacted the minister's office, which in turn called the RCMP to inquire about Roscoe. The RCMP alerted the government about two articles Roscoe had written. One of them, which had appeared in *The Canadian Tribune* eight years earlier, criticized Conservative leader George Drew for nearly getting into a scuffle with CCF leader M.J. Coldwell at a political meeting in Wolfville, Nova Scotia. Roscoe had written that Drew's behaviour showed his "brutality, ignorance and viciousness," and should serve as a warning to Canadians not to elect him. The other, from a *Steelworker and Miner* issue in 1954, was a criticism of U.S. domination of Canada and said that only the Communist Party had asked "the people of Canada to wake up and do something before we find ourselves completely hogtied by Yankee business."[6]

It is not clear from Roscoe's security files whether the Canadian government attempted to help Roscoe gain entry to the United States, but he was granted a special hearing by U.S. authorities in Detroit in January 1957. Going into the hearing, Roscoe didn't expect to win. He had decided not to accept any sort of conditional entry that would require him to report to police while in the country. He wrote his son Dick and daughter-in-law Becky: "I do not have to live under fascism, even for a week."[7]

Following the hearing Roscoe was told that he was still denied entry to the United States, though no reason was given. He tried for more than three years to obtain his records from U.S. Immigration, but met with one delay after another. He applied for admittance again in 1959, appealing to External Affairs for help, but he was never again permitted entry.

From then on, whenever he wanted to see family members in Detroit, Roscoe had to take the train to Windsor and stay in a hotel, where he'd have the relatives come to him. Roscoe was worried that his trouble with U.S. authorities might lead to difficulties for the family members in Detroit. He knew that the damaging information had been obtained by the RCMP and then passed on to the U.S. government. "I do not believe that the government or its police have any legal right to keep such a record on a law-abiding and hard working citizen," he wrote in his autobiography. If he had been able to afford it, he would have gone to court to have the RCMP's file on him destroyed.

❧

In the late 1950s, after Margaret had died, Roscoe became involved in an affair that severely tested his relationship with his family. Manchester Robertson Allison Ltd., Saint John's leading department store, hired Roscoe on a part-time basis in 1954 to provide advice to its gardening customers. Roscoe had just published *Green Thumbs*, he was making appearances on television to talk about gardening, and hundreds of people were asking for his help and advice. At the store, where he returned each spring to give his talks, Roscoe worked with a woman named Helen and took an immediate liking to her.[8] He began secretly—in his words—to carry a torch for her. Roscoe and Helen worked together for four years and in the spring of 1958 a romance developed. Roscoe wrote Herman that "this is not a casual intrigue" and that Helen was "really beautiful and great fun."[9] To be close to her, Roscoe began spending as much time as possible in Saint John.

When he told his immediate family members about the relationship, they became concerned. They discovered that Helen was not only married but also forty-one-years old. Roscoe was seventy-one. Roscoe wanted to bring her to Centreville for a visit, but Rosa adamantly opposed the idea.[10] When Roscoe's next garden book, *Roses for Canadian Gardens*, was published in the summer of 1959, it contained the dedication: "This book is especially dedicated to H.F.P. with all my love and best wishes for a long and happy life. She is a very special person to me and though her name may not appear here she will know who is meant and none other need know, which is as it should be." Dick responded by tearing the dedication page out of the book.

Roscoe was upset about his family's negative reaction and accused Rosa of not wanting him to have anyone to replace Margaret. Rosa replied that it was the particular circumstances she objected to. If he wanted to marry a single woman his own age, that would be fine with her. By the fall of 1959 the affair was slowing down and marriage seemed to be drifting out of the picture. Although Roscoe indicated that Helen couldn't get a divorce, family members speculated that she simply got cold feet when Roscoe became too deeply involved. The relationship soon ended, and for several months Roscoe behaved coolly toward the family members who had been critical of his intentions.

❧

During the time that Roscoe had been infatuated with Helen, the nursery was in a state of crisis. It had become famous for its huge spring crop of pansy plants, a tradition that Roscoe had started when the business first opened in 1924. In 1955 the Fillmores grew 180,000 plants in one field, which they believed to be the largest crop ever in North America. The vast display of pansies, which was featured in a colour spread in *Weekend Magazine*, lured as many as three thousand customers to Centreville on "Pansy Sunday" each spring.[11] The nursery's pick-up truck—which had carried

Roscoe's scrawled message during the war about not buying German or Japanese goods—now sported a bright yellow pansy.

The nursery had always struggled financially, just managing to scrape by and growing deeper into debt in the process. Then it had a disastrous experience with a highly touted new chemical, amino triazole. In fall 1956, after succumbing to a pitch from the giant company North American Cyanamid to use amino triazole as a weed killer for the pansy fields, the family found a full two-thirds of its crop damaged. It turned out that the company had carried out only limited testing on the new chemical and had never tested it on pansies. The corporation refused to accept responsibility for the damages, so the Fillmores launched a lawsuit. When the case was finally settled in February 1958, the nursery was awarded $12,750 plus $2,500 in legal costs. Even so, the award fell several thousand dollars short of covering the total loss caused by the crop destruction and the court case.

Meanwhile the nursery's financial situation remained precarious, despite a move in the spring of 1958 to a new and better location on the main Halifax-Yarmouth highway, just west of Kentville. The family, Roscoe included, was sinking all its resources into the business. The bank was insisting that someone with more financial expertise be brought into the company and was refusing to extend the necessary credit. The nursery's financial situation was eased in February 1958, when the Nova Scotia Industrial Loan Board authorized a loan of $20,000 for the purchase of the property in Kentville. But the frigid winter of 1958-59 saw the heaviest winter-kill in the history of the nursery, and to make matters worse the 120,000-pansy crop was hit by disease.

In July 1960 the Nova Scotia Industrial Loan Board turned down a detailed request for additional help to see the nursery through the crisis.[12] Though the decision was not surprising from a strictly business perspective—given the growing debt and continuing problems with the crop—the Fillmores always suspected that Frank and Roscoe's well-known politics might also have come into play. Roscoe certainly thought so, because he responded with a sharp letter to the board, pointing out that its previous loan had been fully secured and that in not providing an additional loan the board would have to take responsibility for the nursery's failure. Roscoe said that he had recently learned that the board had made a much larger loan to a German Nazi who had fought against Canada during World War II. "My memory is not so short as yours," he wrote. "I assure you gentlemen that I deeply resent your attitudes and actions." He signed the letter, "Yours with complete disrespect."[13]

The nursery went bankrupt in March 1961. While the Fillmores were deeply hurt by the liquidation of their business, no one was ruined financially.

Roscoe lost some of his savings, but he had enough income from his DAR and government pensions to live comfortably. Frank took a job in Truro as a reporter with *The Chronicle-Herald*, specializing in agriculture and horticulture.

❧

The tension caused by the demise of the nursery had its toll. When he was seventy-three years old, Roscoe had serious health problems for the first time. In November 1960, after a period of not feeling well, he went to the doctor and was immediately admitted to hospital. According to the doctor, Roscoe had suffered two coronary thrombosis attacks during the previous four or five years. There had been serious damage and scar tissue had enlarged his heart. Roscoe might not have been helped by a little private medication he had been taking on his own for over a year: He had read in a Russian publication that a small quantity of liquid arsenic taken on a regular basis would extend a person's life. The doctor also found a stomach ulcer that had been there for at least twenty years. Roscoe was sent home from the hospital, put on a special diet, and told to rest. He reluctantly cancelled the six-week trip he'd booked on a passenger freighter to Cuba.

Even when he recovered, Roscoe was unable to keep up the pace he had maintained before his visit to hospital. He tired easily and soon found that he didn't have the energy to continue writing books. He had written most of a manuscript for a fifth gardening book but never finished the project. He had also been working on his autobiography, but had to abandon it after writing only about fifteen thousand words.

Roscoe did have enough energy to write letters to a wide range of relatives and friends, often sending them articles he clipped from the many publications he monitored on a regular basis. He also continued to write letters to the editor, sending them off regularly to *The Chronicle-Herald*, *The Family Herald*, and *The Canadian Tribune*, among others. Although more of Roscoe's letters appeared in *The Chronicle-Herald* than anywhere else, the paper refused to print many of his offerings and often extensively edited the ones it did accept.

The letters to *The Chronicle Herald* sometimes drew harsh responses. Roscoe seldom took notice of the criticisms, but an unsigned letter that appeared in the paper in the early 1960s especially offended him. It offered the somewhat standard comeback that if Roscoe didn't like the way things were in Canada he should go live in Russia. Roscoe retorted that he was seventy-four years old and the one thing he was most proud of was that he had been a dissenter all his life.

> I have read history and learned that without dissenters man would have made no progress. The satisfied and self-complacent are vegetables, nothing more. These are today's greatest problems because they are prepared to follow blind leaders into nuclear cremation and the end of the human race....
>
> There are no infallible or sacrosanct politicians or governments. It is the right and duty of all citizens to criticize those who would lead us, whether they are leaders of our choice or self-appointed. I was in Russia almost all of 1923 as an agricultural specialist. Since that year I have never felt any doubt that Russia would establish socialism. Since it was an experiment, something that had never been done before, it has had its ups and downs. Glaring mistakes have been made as well as outstanding accomplishments achieved....
>
> I do not intend to leave Canada for Russia or any other country. This is my home and I shall spend here the short time left me, criticizing, always criticizing complacency, superstition, injustice. This is my right and duty as I see it.[14]

❧

One of Roscoe's favourite activities when he was in his seventies was to drive his battered Volkswagen Beetle the ninety kilometres to Dartmouth, where Frank and Irene lived. Roscoe was notorious in the family for his bad driving. Rosa would cringe when she saw him leave the yard in Centreville—he didn't look either way as he pulled onto the highway—and she wouldn't relax until she heard he had arrived in Dartmouth. After he suffered a mild stroke in 1964, family members became more and more uneasy about him making these trips. His speech was slightly affected and he walked with the help of a cane.

With the troubled days of the nursery behind them, Roscoe and Frank got along better. Roscoe was proud of the impact Frank was having as a social activist through his journalism. Frank hadn't taken up journalism as a profession until he was thirty-six, but his stinging criticisms helped to bring improvements in Nova Scotia laws concerning the poor, prison reform, housing, and health care.

Roscoe became an enthusiastic supporter of a group of singers that Frank and Irene sponsored and nurtured, the "Reddick Sisters." Their real names were Ruddick and their father, Maurice Ruddick, had gained fame as the "Singing Miner" during the 1956 Springhill mine disaster. After they moved to Dartmouth, and following considerable music training, the girls—Sylvia, nineteen, Valerie, eighteen, and Ellen, sixteen—sang on national television shows such as "Music Hop" and "Don Messer's Jubilee."

Roscoe's interest in the Reddicks seemed to give him new life and energy. The girls' grandparents were black and, ever conscious of the value of education in political and social issues, Roscoe talked to them on racism in Nova Scotia and South Africa and on black history in North America. He sometimes went to the "Black Panther," a coffee house that Frank and Irene set up in Dartmouth as a platform for the singers. At seventy-six, with his white hair and his cane, he would sit listening to music and debating the Vietnam War and the American civil rights movement with the young customers. "I am sure you would laugh if you saw me at the coffee house," he wrote Dick and Becky. "I'm quite sure my sainted Grandmother must be in distress!"[15] Eventually, after finding their new life too much of a strain, the Reddick Sisters gave up their plans for a full-time music career.

In 1967 Roscoe received one of his last and most treasured accolades for his contributions to gardening. As a part of Canada's Centennial celebrations, the Canadian and Ontario Nursery Trades Associations published the *Canadian Nurseryman Centennial Yearbook*, a study of the formation and growth of the horticultural industry in the country. In a section entitled "Great Canadians in Horticulture," Roscoe was honoured as one of twenty-four people who had made an outstanding contribution to the development of horticulture in Canada—the only person to be so honoured from the Atlantic region.

That same year, though, he suffered a terrible loss: His daughter Allie died, at age forty-five, leaving behind her husband, who was in the Canadian Air Force, and two small children. Roscoe was shocked that he had outlived one of his children, and Allie had always had a special place in Roscoe's heart. She was only eighteen months old when Roscoe had gone off to Russia, and he had felt at the time that he might never return to see her grow up. Roscoe, badly shaken by Allie's death, and in search of some explanation, had to remind himself that he didn't place his faith in religion. Writing to his granddaughter, Barbara Tarbuck, Roscoe quoted a line written by a socialist he had known many years before, Gerald O'Connell Desmond: "Girls my only angels, men my more than gods, earth holds all I know of heaven or hell."[16]

In January 1968, Roscoe's health began to deteriorate. He was admitted to hospital, and examinations showed stones in his kidneys and bladder as well as an abnormal growth of the prostate. The doctors treated the stones, but further studies revealed that Roscoe had cancer of the prostate. Following an operation, Roscoe was put on hormones in an attempt to control the growth of the cancer. He wrote to friends that he was weak but improving, although he was in fact extremely ill. A radical until the end, he wrote to his long-time friend Dane Parker that he hoped the operation and treatment would give him "a few more months or even years to watch the world go by and perhaps see some of the stinking bastards hanged."[17]

He left hospital and returned home to Centreville. On June 25 there was a federal election, and with Trudeaumania sweeping the nation Roscoe went to the polls and voted for the NDP, even though he felt lukewarm about supporting them. "I don't know that the NDP is worth voting for," he wrote. "Too many professors, lawyers, teachers and outstandingly respectable people. I voted for them but without enthusiasm."[18]

In early November Roscoe received a letter from Tim Buck, asking him to provide information on the early days of the Canadian Communist Party so its history could be written.[19] Roscoe was never able to reply. The condition of his heart became so bad that once again he was admitted to hospital, where he died a week later on November 20, 1968, at the age of eighty-one.

❧

For most of his life Roscoe Fillmore had been obsessed with social and political problems and the solution offered by socialism. It is here that one would expect to find his greatest legacy. Yet his life's work as an activist and an organizer directly influenced only a small number of people. He never held public office, and—despite the arduous work of the RCMP—those in positions of authority seem to have paid little attention to what he said. Nevertheless, Roscoe and others like him helped to introduce a new political and social philosophy to the Maritimes. He was deeply involved in the birth and, ironically, the demise of radical socialism in the region.

If Roscoe left a legacy for socialists, it is probably the message that despite their small numbers and the loneliness and difficulty of being socialists, they should remain true to the cause. Roscoe was determined that people should never give up fighting for their ideals. He never abandoned his own faith that some day a co-operative form of society would replace capitalism in Canada and throughout the world.

Sadly, Roscoe died just months before my father Frank Fillmore and I launched *The People*, the first of three small newspapers published in Nova Scotia. Later, *The 4th Estate* and *The Scotian Journalist*, each in their own way, reflected some of the indignation and socialist morality that had been Roscoe's trademark for so many years.

Roscoe's most tangible legacy is his contribution to horticulture. His popular books helped thousands of amateur gardeners. He gained national recognition for his propagation work, and the success of his experiments ensured that hardy species of plants were able to survive in the harsh climate of the Maritimes. To this day, the plants he helped to create—such as the rhododendrons at Grand Pre Memorial Park and the Kentville Experimental Farm—constitute a visible reminder of Roscoe's dedication to his profession.

In July 1978, on the ninetieth anniversary of Roscoe's birth, the first Roscoe Fillmore Memorial Picnic was held at the home of Kaye and Charlie Murray in Lower Sackville. More than two hundred friends and admirers, many of them socialists of one variety or another, sat in the yard of the Murray home, amid Roscoe's plants and flowers, recalling what they could of Roscoe and talking about the difficult times facing socialism. Delivering a tribute to Roscoe at the first picnic, Dane Parker read Ken Leslie's poem, "There Is A Man Within."

> There is a man within, a sure one.
> He having taken your heart will hold it ever,
> will hug and hold his treasure ever and ever.
> You may wander and lose yourself, you may return,
> you may forget him, you may betray this lover,
> but he will never mislay the heart you have given.
> He will hold his treasure forever and ever.[20]

Notes

ONE In the Backwoods of New Brunswick

1 James Hannay, *History of New Brunswick* (Saint John: John A. Bowes, 1909); Graeme Wynn, *Timber Colony: A Historical Geography of Early Nineteenth Century New Brunswick* (Toronto: University of Toronto Press, 1981); Hugh G. Thorburn, *Politics in New Brunswick* (Toronto: University of Toronto Press, 1961), Chapter 1; W.C. Milner, "History of Albert County," unpublished working copy, New Brunswick Museum, Saint John; Albert W. Smith, *Essay on the History and Resources of Albert County* (Fredericton: The Journals of the House of Assembly, 1907), Appendix G; Orland R. Atkinson, "History and Resources of Albert County," *The Albert Journal*, July 18, 25, August 1, 8, and 15, 1906; Alfred C. Barbour, "Essay on Riverside, Albert County," *Moncton Daily Times*, December 3, 1910.

2 Charles L. Fillmore, *So Soon Forgotten: Three Thousand Fillmores* (Rutland, Vt.: Daamen, 1984), p.243.

3 Roscoe Fillmore, "Autobiography," unpublished manuscript, Dalhousie University Archives, Halifax, 1959–60. Unless otherwise noted, all unattributed quotes from Roscoe Fillmore are from this source.

4 Margaret Conrad, Toni Laidlaw, and Donna Smyth, *No Place Like Home: Diaries and Letters of Nova Scotia Women, 1771–1938* (Halifax: Formac Publishing, 1988), p.14.

5 See Charles W. Deweese, "Church Covenants and Church Discipline among Baptists in the Maritime Provinces," in *Repent and Believe: The Baptist Experience in Maritime Canada*, ed. Barry M. Moody (Hantsport, N.S.: Lancelot Press for Acadian Divinity College, 1980), pp.27–45. From 1848 to 1877, during which time John and Elizabeth were active in the Baptist church, many hundreds of

church members in eastern New Brunswick were publicly censured for their behaviour, and 893 people were excluded from the church entirely.

6 *The Maple Leaf* (Albert, N.B.), November 8, 1888.

7 *The Daily Times* (Moncton), December 3, 1910. Essay on Riverside, Albert County, by Alfred C. Barbour.

8 Thorburn, *Politics in New Brunswick*, pp.6–10; A.R.M. Lower, *The North American Assault on the Canadian Forest* (New York: Yale University Press, 1938), p.166; Wynn, *Timber Colony*, Chapter 3.

9 Herman Fillmore, "Fillmore–Tingley Family History," unpublished manuscript, Chapter 1, p.8.

10 T.W. Acheson, "The National Policy and the Industrialization of the Maritimes, 1880–1910," in *Industrialization and Underdevelopment in the Maritimes, 1880–1930*, ed. T.W. Acheson, David Frank, and James Frost (Toronto: Garamond Press, 1985). According to Acheson, the Maritimes had benefited substantially from the National Policy by 1885. With less than one-fifth of the population of Canada, the region contained eight of the nation's twenty-three cotton mills, three of five sugar refineries, two of seven rope factories, one of three glass works, both of the Canadian steel mills, and six of the nation's rolling mills.

11 Figures cited in Alan A. Brookes, "Out-Migration from the Maritimes Provinces, 1860–1900: Some Preliminary Considerations," *Acadiensis*, Spring 1976, p.29; Patricia A. Thornton, "The Problem of Out-Migration from Atlantic Canada, 1871–1921: A New Look," *Acadiensis*, Autumn 1985; David Alexander, "Economic Growth in the Atlantic Region, 1880 to 1940," *Acadiensis*, Autumn 1978.

12 *The Maple Leaf*, October 18, 1894.

13 *The Maple Leaf*, November 22, 1894.

TWO The Germination of a Radical

1 Nancy Jon Colpitts, "Alma, New Brunswick and the Twentieth Century Crisis of Readjustment: Sawmilling Community to National Park," M.A. thesis, Dalhousie University, Halifax, 1983, p.58.

2 *McAlpine's Directory*, New Brunswick, 1903, pp.453–457.

3 S.A. Saunders, *The Economic History of the Maritime Provinces: A Study Prepared for the Royal Commission on Dominion-Provincial Relations* (Ottawa, 1939), pp.4–5, Appendix Table No. 2.

4 Alfred C. Barbour, *The Daily Times*, December 3, 1910.

5 Robert Babcock, "Economic Development in Portland, Me. and Saint John, N.B. during the Age of Iron and Steam, 1850–1914," *The American Review of Canadian Studies*, Vol.IX, No.1 (1979), p.23.

6 Roscoe Fillmore, "Socialism," Riverside Consolidated School essay, Riverside, N.B., 1906, in Dalhousie University Archives, Halifax.

THREE The Great Harvest Excursion

1 "Harvest Special," *Current Account, Personnel Magazine of the Canadian Bank of Commerce*, September 1960, p.5.

2 John Herd Thompson, "Bringing in the Sheaves: The Harvest Excursionists, 1890–1929," *The Canadian Historical Review*, LIX.4. (1978), pp.485–486.

3 W.J.C. Cherwinski, "The Incredible Harvest Excursion of 1908," *Labour/Le Travailleur*, Spring 1980, p.58.

4 W. Kaye Lamb, *Railroads of America: History of the Canadian Pacific Railway* (New York: Macmillan Publishing, 1977), Chapter 19.

5 Edmund Bradwin, *The Bunkhouse Men: A Study of Work and Play in the Camps of Canada 1903–1914* (New York: Columbia University Press, 1928), pp.44–45.

6 Donald Avery, *'Dangerous Foreigners': European Immigrant Workers and Labour Radicalism in Canada, 1896–1932* (Toronto: McClelland and Stewart, 1979), pp.52–56.

7 Ibid., Chapter III.

8 The story of Roscoe's train experience is based on Roscoe's account in *Cotton's Weekly* (Cowansville, Que.), November 18, 1909, and Roscoe Fillmore's autobiography.

9 Cited in *The Best of Bob Edwards*, ed. Hugh A. Dempsey (Edmonton: Hurtig Publishers, 1975), p.56.

10 Telephone interview with Franklin Rosemont, Charles Kerr and Company, Chicago, January 3, 1990.

11 *The Western Clarion* (Vancouver), December 21, 1907.

12 Melvyn Dubofsky, "The Origins of Western Working Class Radicalism," in *The American Labor Movement*, ed. David Brody (New York: Harper and Row, 1971), pp.83–99.

13 A. Ross McCormack, *Reformers, Rebels, and Revolutionaries: The Western Canadian Radical Movement 1899–1919* (Toronto: University of Toronto Press, 1977) Chapter 1.

14 Ibid., Chapter 4.

15 *Cotton's Weekly*, October 28, 1909.

FOUR Socialism for the Maritimes

1 *Cotton's Weekly*, January 28, 1909.

2 *Periodical Bulletin of the International Socialist Bureau*, cited in *Cotton's Weekly*, December 2, 1909.

3 The publications Roscoe read included *The Western Clarion* (Vancouver), *New York Call*, *International Socialist Review* (Chicago), and *Coming Nation* and *Appeal to Reason*, both published in Girard, Kansas.

4 Interview with Clara Fillmore Reid, Detroit, July 30, 1988.

5 G.H. Allaby, "New Brunswick Prophets of Radicalism: 1890–1914," M.A. thesis, University of New Brunswick, 1972.

6 James K. Chapman, "Henry Harvey Stuart (1873–1952): New Brunswick Reformer," *Acadiensis*, No. 5 (1972).
7 Thorburn, *Politics In New Brunswick*, p.55.
8 *The Western Clarion*, July 4, 1908.
9 Interview with Mabel Fillmore Gross, Detroit, July 30, 1988.
10 Roscoe Fillmore letter to H.H. Stuart, June 8, 1908, H.H. Stuart Collection, Harriet Irving Archives, University of New Brunswick, Fredericton.
11 *The Albert Journal* (Hillsboro, N.B.), April 15, 1908.
12 Roscoe Fillmore letter to H.H. Stuart, July 22, 1909, H.H. Stuart Collection, Harriet Irving Library, University of New Brunswick, Fredericton.
13 *Cotton's Weekly*, March 11, 1909.
14 *The Western Clarion*, May 22, 1909.
15 *Cotton's Weekly*, May 20, 1909.
16 Roscoe Fillmore letter to H.H. Stuart, June 24, 1909, H.H. Stuart Collection, Harriet Irving Library, University of New Brunswick, Fredericton.
17 *Cotton's Weekly*, August 5, 1909.
18 *Cotton's Weekly*, September 23, 30, 1909.
19 *The Western Clarion*, October 30, 1909.
20 David Frank and Nolan Reilly, "The Emergence of the Socialist Movement in the Maritimes, 1899–1916," *Labour/Le Travailleur*, 1979, pp.85–86.
21 Ibid., p.95.

FIVE The Jail Birds of Liberty

1 Melvyn Dubofsky, *We Shall Be All: A History of the Industrial Workers of the World* (New York: Quadrangle/New York Times Book Co., 1969), Chapter 8.
2 *The Daily Telegraph* (Saint John), November 1, 1909.
3 *Cotton's Weekly*, November 18, 1909.
4 William D. Haywood letter to Roscoe Fillmore, November 12, 1909, property of Barbara Tarbuck.
5 *The Daily Telegraph*, November 11, 1909.
6 *The Eastern Labor News* (Moncton), November 27, 1909.
7 Dubofsky, *We Shall Be All*, pp.241–242.
8 *The Daily Times*, November 27, 1909.
9 *Cotton's Weekly*, December 23, 1909.
10 Craig Heron, "Labourism and the Canadian Working Class," *Labour/Le Travail*, Spring 1984, pp.47–50.
11 *The Western Clarion*, November 19, 1910.
12 *The Eastern Labor News*, November 4, 1911.
13 Ibid.
14 *Cotton's Weekly*, April 8, 1909.

15 *Cotton's Weekly*, September 6, 1909.
16 Heron, "Labourism and the Canadian Working Class," p.65.
17 Ian McKay, "Strikes in the Maritimes, 1901–1914," unpublished paper, Dalhousie University, Halifax, 1980, pp.4–5.
18 *The Daily Times*, July 24, 1908.
19 *The Eastern Labor News*, May 15, 1909.
20 *The Eastern Labor News*, May 29, 1909.
21 The information on Amherst draws on the research of historians Nolan Reilly and David Frank, including Reilly, "The Emergence of Class Consciousness in Industrial Nova Scotia: A Study of Amherst, 1891–1925," Ph.D. thesis, Dalhousie University, Halifax, 1983; Reilly, "The Origins of the Amherst General Strike 1880–1919," unpublished paper, Dalhousie University, May 1977; Reilly and Frank, "Emergence of the Socialist Movement in the Maritimes"; and "The Rise and Fall of Busy Amherst," a CBC Radio "Ideas" program prepared by Reilly and produced by Tom MacDonnell.
22 *Amherst Daily News*, July 6, 1909.
23 Roscoe's friends who were members of the Socialist Party of Canada in Amherst included tailors Dan McDonald and George McLeod, who were the principal forces behind the formation of the local chapter of the Journeymen Tailors' Union of America. Two other close friends were unionists John Logan and Tom Godfrey, Scotsmen and moulders who, with their wives, were socialists. There was Arthur MacArthur, originally from Albert County, whom Roscoe remembered from his boyhood. Other prominent socialists included William McInnis, Zabred McLeod, and Hillman Farnell of the car works, Joseph Mitchell of the Victor Woodworking Co., and Thomas Carr from the shoe works.
24 Reilly, "Origins of the Amherst Strike," pp.7–8.
25 *Cotton's Weekly*, August 18, 1910.

SIX Class Struggle in Springhill 1909–11

1 Ian McKay, "Industry, Work and Community in the Cumberland Coalfields, 1848–1927," Ph.D. thesis, Dalhousie University, Halifax, 1983; Ian McKay, "The Realm of Uncertainty: The Experience of Work in the Cumberland Coal Mines, 1873–1927," *Acadiensis*, Fall 1986; Helen S. Goodwin, "Community, Class and Conflict: The 1909–1911 Springhill Coal Strike," Honours thesis, Dalhousie University, Halifax, 1980; Danny Moore, "The 1909 Strike in the Nova Scotia Coal Fields," unpublished M.A. research paper, Carleton University, Ottawa, 1977.
2 McKay, "Realm of Uncertainty," p.42.
3 *Cotton's Weekly*, May 20, 1909.
4 McKay, "Realm of Uncertainty," p.3.
5 Bertha J. Campbell et al., *Springhill: Our Good Heritage* (Springhill Heritage

Group, 19898), pp.v–vi; Roger David Brown, *Blood on the Coal: The Story of the Springhill Mining Disasters* (Hantsport, N.S.: Lancelot Press, 1976), pp.78–79.
6 *Amherst Daily News*, August 11, 1909.
7 Jean Heffernan, *The Springhill Record* (1950), cited in Campbell et al., *Springhill*, p.221.
8 *The Halifax Herald*, August 11, 1910.
9 *Cotton's Weekly*, June 30, 1910.
10 Ibid.
11 McKay, "Industry, Work and Community," pp.215–228.
12 *The Eastern Labor News*, November 12, 1910.
13 McKay, "Industry, Work and Community," p.343.
14 *The Halifax Herald*, August 31, 1909.
15 *Cotton's Weekly*, June 9, 1910.
16 *Cotton's Weekly*, June 16, 1910.
17 *The Western Clarion*, June 11, 1910.
18 *Cotton's Weekly*, June 30, 1910.
19 *Cotton's Weekly*, June 23, 1910.
20 *Cotton's Weekly*, September 8, 1910.
21 *Cotton's Weekly*, July 14, 1910.
22 *Cotton's Weekly*, June 30, 1910.
23 *The Eastern Labor News*, July 30, 1910.
24 Ibid.
25 *Cotton's Weekly*, August 18, 1910.
26 *Cotton's Weekly*, July 14, 1910.
27 *Cotton's Weekly*, August 26, 1910.
28 Robert McIntosh, "The Boys in the Nova Scotia Coal Mines: 1873–1923," *Acadiensis*, Spring 1987; McIntosh, "Grotesque Faces and Figures: Child Labourers and Coal Mining Technology in Victorian Nova Scotia," *Scientia Canadensis*, Fall/Winter 1988, p.97; McKay, "Realm of Uncertainty," pp.24–33.
29 *Cotton's Weekly*, January 26, March 16, 1911.
30 McKay, "Industry, Work and Community," pp.796,849–851.
31 The *Amherst Daily News*, June 3, 1911.

SEVEN "Yours in Revolt, Roscoe Fillmore"

1 *The Western Clarion*, November 19, 1910.
2 *The Daily Herald* (Calgary), January 7, 1908; Robert H. Babcock, "Saint John Longshoremen During the Rise of Canada's Winter Port, 1895–1922," *Labour/Le Travail*, Spring 1990, p.32.
3 *The Western Clarion*, November 18, 1911.
4 Roscoe Fillmore, "Strikes and Socialism in Eastern Canada," *The International Socialist Review*, April 1910, p.890.

5 *Cotton's Weekly*, June 17, 1909.
6 For more on this issue, see Linda Kealey, "Canadian Socialism and the Woman Question, 1900–1914," *Labour/Le Travail*, Vol.13 (Spring 1984); Linda Kealey and Joan Sangster, eds., *Beyond the Vote: Canadian Women and Politics* (Toronto: University of Toronto Press, 1989); Joan Sangster, *Dreams of Equality: Women On The Canadian Left, 1920–1950* (Toronto: McClelland and Stewart, 1989).
7 *Cotton's Weekly*, August 19, 1909.
8 Frank and Reilly, "Emergence of the Socialist Movement in the Maritimes," p.106.
9 *Cotton's Weekly*, July 1, 1909.
10 *Cotton's Weekly*, September 23, 1909.
11 Correspondence to the author from Desmond Morton, Toronto, July 10, 1989.
12 *Cotton's Weekly*, December 2, 1909.
13 Registry of Church Members, Hopewell United Baptist Church, Albert, N.B., Atlantic Baptist Historical Collection, Vaughan Memorial Library, Acadia University, Wolfville, N.S., entry for June 14, 1911.
14 Helena Sheehan, *Marxism and the Philosophy of Science: A Critical History*, Vol. One, *The First Hundred Years* (London: Humanities Press, 1985), p.23.
15 Ibid., p.45.
16 *Cotton's Weekly*, August 26, 1909.
17 *Cotton's Weekly*, February 17, 1910.
18 McCormack, *Reformers, Rebels, and Revolutionaries*, p.71.
19 *Amherst Daily News*, September 15, 21, 1911.
20 *The Western Clarion*, October 1911.
21 *The Western Clarion*, November 25, 1911.
22 *The Western Clarion*, March 29, 1913.
23 Roscoe Fillmore, "Keep the Issue Clear," *The International Socialist Review*, September 1914, p.398.

EIGHT Apple Farming, War, and a Revolution

1 *The Daily Gleaner* (Fredericton), April 26, 1910, gives the company's original plans, which included dividing the huge farm into smaller pieces to be farmed by tenant farm families brought over from England. This plan never materialized.
2 Barry Grant, a Fredericton historian working on the history of Sunbury County, provided background in a telephone interview, July 22, 1990.
3 *The Albert Journal*, March 6, 1912.
4 *The Graves Papers*, Vol.V, Part II, p.128, Provincial Archives of New Brunswick, Fredericton.
5 Arthur T. Doyle, *Front Benches and Back Rooms: A Story of Corruption, Muckraking, Raw Partisanship and Intrigue in New Brunswick* (Toronto: Green Tree Publishing, 1976), Chapter 2.
6 Ibid., Chapter 4.

7 Linda Kealey, "Sophie," *New Maritimes*, November 1987.
8 Fillmore, "Keep the Issue Clear," p.401.
9 *Saint John Globe*, August 5, 1914.
10 *The Western Clarion*, September 1914.
11 Roscoe Fillmore, "How to Build Up the Socialist Movement," *The International Socialist Review*, 1915, p.615.
12 Ibid., p.616.
13 Herman Fillmore, unpublished memoirs, p.172.
14 Fillmore, "How to Build Up the Socialist Movement," p.616.
15 Fillmore, "Keep the Issue Clear," p.399.
16 *The Western Clarion*, June 1915.
17 Doyle, *Front Benches and Back Rooms*, p.85.
18 *Saint John Globe*, January 12, 1916.
19 *Saint John Globe*, January 11, 1916.
20 *Saint John Globe*, January 20, 1916.
21 *The Daily Mail*, Fredericton, February 12, 1916.
22 Fillmore, "How to Build Up The Socialist Movement," p.616.
23 *The Daily Gleaner*, March 16, 1917.
24 The date of the Russian Revolution, according to our calendar, was November 7. But Czarist Russia was still using the Julian calendar, which was thirteen days behind ours, so the revolution occurred in October and was called the Great October Revolution.
25 Hans Werner, "And What, Exactly, Was Leon Trotsky Doing in Nova Scotia in 1917?" *Saturday Night*, August 1974.
26 Walter Lippmann and C. Merz, "A Test of The News," *New Republic*, August 4, 1920, p.10.
27 Phillip Knightley, *The First Casualty: From the Crimea to Vietnam: The War Correspondent as Hero, Propagandist, and Myth Maker* (New York: Harcourt Brace Jovanovich, 1982), p.146.
28 *Amherst Daily News*, February 24, 1919.

NINE Red Scare: The Socialist Party of Canada

1 Roscoe Fillmore letter to R.B. Russell, November 9, 1918, R.B. Russell Collection (MG10 A14–1, Box 3, Item 13), Manitoba Provincial Archives (MPA), Winnipeg.
2 David J. Bercuson, *Fools and Wise Men: Rise and Fall of the One Big Union* (Toronto: McGraw-Hill Ryerson, 1978), pp.58–59.
3 Gustavus Myers, *History of Canadian Wealth* (Toronto: James, Lewis and Samuel, 1973), p.i.
4 Gregory S. Kealey, "1919: The Canadian Labour Revolt," *Labour/Le Travail*, Spring 1984, p.11.

5 Roscoe Fillmore letter to Socialist Party of Canada, October 8, 1918, R.B. Russell Collection, MPA, Winnipeg.
6 Letter from Fred Thompson to John Bell, August 9, 1976, Labour History Collection, Dalhousie University Archives, Halifax; Suzanne Morton, "Labourism and Economic Action: The Halifax Shipyards Strike of 1920," *Labour/Le Travail*, Fall 1988.
7 Other men Roscoe interested in socialism were labourer William McFadyen, who lived near Burton; Ernest Camp of Sunbury County; George Danby of Fredericton Junction; and Silas Ward of Oromocto.
8 *The Daily Gleaner*, April 5, 1923.
9 Roscoe Fillmore letter to R.B. Russell, December 5, 1918, R.B. Russell Collection, MPA, Winnipeg.
10 Roscoe Fillmore letter to Socialist Party of Canada, February 27, 1919, R.B. Russell Collection, MPA, Winnipeg.
11 *The Daily Gleaner*, July 10, 1919.
12 Ibid.
13 *The Red Flag* (Vancouver), April 26, May 3, 1919.
14 Roscoe Fillmore letter to C. Stephenson, February 25, 1919, R.B. Russell Collection, MPA, Winnipeg.
15 Ibid.
16 The loose-knit group included Dan McDonald of the tailoring shops union (he had helped Roscoe and the socialists take over the Cumberland Labor Party in 1909) and four members of the car-works union: William McInnis, Zabred McLeod, William Godfrey, and Joseph Mitchell. New members included Karl Rockwood, W. Jones, and Alfred Barton, who was secretary of the Amherst Federation of Labor.
17 Roscoe Fillmore letter to the Socialist Party of Canada, February 27, 1919, R.B. Russell Collection, MPA, Winnipeg.
18 *The Labour Gazette* (Ottawa), July, 1919, p.834.
19 Nolan Reilly, "The General Strike in Amherst, Nova Scotia, 1919," *Acadiensis*, No.9 (1980); Nolan Reilly, "Notes on the Amherst General Strike and the One Big Union," *Bulletin of the Committee on Canadian Labour History*, No.3 (Spring 1977); and Reilly, "The Emergence of Class Consciousness in Industrial Nova Scotia."
20 Norman Penner, "How the RCMP Got Where It Is," in *RCMP vs The People: Inside Canada's Security Service*, by Edward Mann and John Alan Lee (Don Mills, Ont.: General Publishing, 1979), p.111.
21 Ibid., p.115.
22 *Western Clarion*, February 15, 1921.
23 Norman Penner, *The Canadian Left: A Critical Analysis* (Scarborough, Ont.: Prentice-Hall, 1977), pp.66–68.

TEN A Journey to the "Promised Land"

1 Dubofsky, *We Shall Be All*, p.239.
2 Lippmann and Merz, "Test of the News," p.10.
3 Dubofsky, *We Shall Be All*, Chapters 17, 18.
4 William D. Haywood, *The Autobiography of Big Bill Haywood* (New York: International Publishers, 1929), Chapter 24.
5 William D. Haywood letter to Roscoe Fillmore, October 11, 1922, property of Barbara Tarbuck, Los Angeles.
6 H.N. Brailsford, *The New York Call*, September 15, 1921.
7 William D. Haywood letter to Roscoe Fillmore, October 11, 1922, property of Barbara Tarbuck, Los Angeles.
8 Interview with Rosa Fillmore Skinner, New Minas, N.S., August 22, 1988.
9 *The Daily Gleaner*, April 5, 1923.
10 J.P. Morray, *Project Kuzbas: American Workers in Siberia (1921–1926)* (New York: International Publishers, 1983) p.62.
11 Roscoe Fillmore letter to Margaret Munroe Fillmore, April 8, 1923, Dalhousie University Archives, Halifax.
12 Roscoe Fillmore letter to Margaret Munroe Fillmore, April 18, 1923, Dalhousie University Archives, Halifax.
13 Ibid.
14 Ibid.
15 Roscoe Fillmore letter to Margaret Munroe Fillmore, April 19, 1923, Dalhousie University Archives, Halifax.
16 Roscoe Fillmore letter to Margaret Munroe Fillmore, April 27, 1923, Dalhousie University Archives, Halifax.
17 Ibid.
18 *Boston Sunday Advertiser*, November 14, 1920. This and other newspaper clippings were found among Roscoe's books and papers.
19 Henry G. Alsberg, *The New York Call*, June 16, 1921.
20 *One Big Union Bulletin* (Winnipeg), October 11, 1923.
21 *One Big Union Bulletin*, September 13, 1923.
22 Ibid.
23 Roscoe Fillmore letter to Margaret Munroe Fillmore, May 3, 1923, Dalhousie University Archives, Halifax.

ELEVEN Siberia: A Gardener for the Revolution

1 Roscoe Fillmore letter to Margaret Munroe Fillmore, June 23, 1923, Dalhousie University Archives, Halifax.
2 Ibid.
3 Ibid.

4 Roscoe Fillmore letter to Margaret Munroe Fillmore, June 15, 1923, Dalhousie University Archives, Halifax.

5 Ibid.

6 Fred Thompson worked for the IWW for almost sixty-five years. He organized miners in Montana and Colorado in the 1920s and autoworkers in Detroit in the 1930s; he became editor of the union's newspaper and later the IWW's historian. Thompson eventually became president of Charles H. Kerr Publishing Company of Chicago—the same publisher that Roscoe had peddled books for in Calgary in 1907 when he was unemployed.

7 Roscoe Fillmore letter to Margaret Munroe Fillmore, June 23, 1923, Dalhousie University Archives, Halifax.

8 Roscoe Fillmore letter to Ruth Fillmore, July 8, 1923, Dalhousie University Archives, Halifax.

9 Roscoe Fillmore letter to Richard Fillmore, July 3, 1923, Dalhousie University Archives, Halifax.

10 Roscoe Fillmore letter to Margaret Munroe Fillmore, June 15, 1923, Dalhousie University Archives, Halifax.

11 Roscoe Fillmore letter to Margaret Munroe Fillmore, July 31, 1923, Dalhousie University Archives, Halifax.

12 Roscoe Fillmore letter to Margaret Munroe Fillmore, August 9, 1923, Dalhousie University Archives, Halifax.

13 Roscoe Fillmore letter to Margaret Munroe Fillmore, August 20, 1923, Dalhousie University Archives, Halifax.

14 Roscoe Fillmore letter to Margaret Munroe Fillmore, September 6, 1923, Dalhousie University Archives, Halifax.

15 Roscoe Fillmore letter to Margaret Munroe Fillmore, September 15, 1923, Dalhousie University Archives, Halifax.

16 Dubofsky, *We Shall Be All*, pp.460–461.

17 Walter J. Lemon letter to Roscoe Fillmore, August 1924, Dalhousie University Archives, Halifax.

18 Ruth Kennell, "Kuzbas in 1924," *The Nation*, November 26, 1924.

19 Morray, *Project Kuzbas*, Chapter 9.

20 *The Maritime Labor Herald*, February 2, 1924.

TWELVE New Life and Politics in the Annapolis Valley

1 *The Morning Chronicle* (Halifax), January 1, 1924, p.23.

2 Roscoe Fillmore, "Propagation in the Small Nursery," address to the International Plant Propagators Society, Cleveland, Ohio, December 2, 1954.

3 *Place Names and Places of Nova Scotia*, Introduction by C.B. Ferguson (Halifax:

Public Archives of Nova Scotia, 1967), p.118; A.W.H. Eaton, *The History of Kings County* (Belleville, Ont.: Mika Studio, 1972).

4 "Report of the Royal Commission Appointed to Investigate the Fruit Industry of Nova Scotia (Halifax, 1930), cited in Margaret Conrad, "Apple Blossom Time in the Annapolis Valley 1880–1957," *Acadiensis*, Vol.IX, No.2 (Spring 1980).

5 *The Advertiser* (Kentville, N.S.), June 18, 1926.

6 Fillmore, "Propagation in the Small Nursery."

7 Roscoe Fillmore letter to Dane Parker, Halifax, January 14, 1948.

8 Interview with Rosa Fillmore Skinner, New Minas, N.S., August 22, 1988.

9 *The Advertiser*, August 25, 1927.

10 *The Advertiser*, September 1, 1927.

11 *Charles Wm MacDonald: Seaman, Labourer, Artist, Manufacturer (1874–1967)* (Halifax: The Art Gallery of Nova Scotia, undated).

12 Susan Perly, "We Bury Our Poets. Kenneth Leslie: A Homesick Bluenoser," *Canadian Forum*, June 1975; Burris Devanney, "Kenneth Leslie: A Biographical Introduction," *Canadian Poetry*, No.5 (Fall/Winter 1979), p.83.

13 Suzanne Morton, "Labourism and Independent Labour Politics in Halifax, 1919–1926," M.A. thesis, Dalhousie University, Halifax, 1986; "Joe Wallace: How I Began," excerpts from a 1975 interview by Alan Safarik and Dorothy Livesay, condensed by Gary Burrill, *New Maritimes*, April 1987; John Robert Colombo, "Joe Wallace: A Banned Poet," *The Varsity* (University of Toronto), February 10, 1960.

14 Ian Angus, *Canadian Bolsheviks: Early Years of the Communist Party of Canada* (Montreal: Vanguard Publications, 1981), pp.244–246.

15 Norman Penner, *Canadian Communism: The Stalin Years and Beyond* (Toronto: Methuen, 1988), pp.1,11.

16 Ibid., p.244.

17 Penner, *The Canadian Left*, p.80.

18 Interview with Dane Parker, Halifax, August 20, 1988.

THIRTEEN The Dirty Thirties: The Poor Shall Want

1 Lorne Brown, *When Freedom Was Lost* (Montreal: Black Rose Books, 1987), p.19; *The Halifax Chronicle*, July 9, June 30, 1931.

2 Barbara Roberts, *Whence They Came: Deportation from Canada 1900–1935* (Ottawa: University of Ottawa Press, 1988), Chapter 7.

3 John Herd Thompson with Allen Seager, *Canada 1922–1939: Decades of Discord* (Toronto: McClelland and Stewart, 1985), pp.228–229.

4 RCMP report from Kentville Detachment to Headquarters, April 28, 1933.

5 *The Steelworker* (Sydney, N.S.), April 4, 1936.

6 Hope McPhee file, Dalhousie University Archives, Halifax.

7 *The Halifax Chronicle*, January 1, 1934.

8 Brown, *When Freedom Was Lost*, p.18.
9 *The Steelworker*, January 12, 1935.
10 Interview with Ruth Marvin, Toronto, September 21, 1988.
11 Interview with Dane Parker, Halifax, August 20, 1988.
12 *New Maritimes*, April 1987; interview with Dane Parker, Halifax, August 20, 1988; John Robert Colombo, "Joe Wallace, A Banned Poet," *The Varsity*, February 10, 1960, p.5.
13 Jim Green, *Against the Tide: The Story of the Canadian Seamen's Union* (Toronto: Progress Books, 1986), p.131.
14 Interview with Dane Parker, Halifax, February 2, 1990.
15 Interview with Dane Parker, Halifax, February 5, 1990.
16 *The Steelworker*, May 4, 1935.
17 E.R. Forbes, "Cutting the Pie into Smaller Pieces: Matching Grants and Relief in the Maritime Provinces during the 1930s," *Acadiensis*, Autumn 1987.
18 *The Halifax Chronicle*, November 24, 1934.
19 *The Halifax Chronicle*, December 22, 1934.
20 *The Steelworker*, November 24, 1934.
21 *The Steelworker*, February 9, 1935.
22 *The Worker* (Toronto), May 24, 1935.
23 Interview with Dane Parker, Halifax, April 29, 1990. Records of the membership of the Communist Party of Canada have been destroyed, but Parker believes that Roscoe was a member of the party's central committee during this period.
24 Mike Earle, "Cape Breton's Radical Weekly Newspapers, 1930–1950," unpublished essay.
25 *The Halifax Chronicle*, November 7, 1934.
26 *The Steelworker*, November 24, 1934.
27 J. Murray Beck, *The Government of Nova Scotia* (Toronto: University of Toronto Press, 1957), p.249.
28 *The Halifax Chronicle*, June 10, 1935.
29 RCMP cross-reference sheet report, October 9, 1935.
30 RCMP report, Sydney subdivision, October 17, 1935.

FOURTEEN Fish Plants and Workers' Rights

1 Thompson with Seager, *Canada 1922–1939*, p.285.
2 Telephone interview with George MacEachern, Glace Bay, N.S., September 22, 1988.
3 *The Steelworker*, April 4, 1936.
4 *The Steelworker*, December 1, 1934.
5 Government of Canada, *Report of the Royal Commission on Price Spreads*, Ottawa, 1935, pp.176,184.
6 *Steelworker and Miner* (Sydney, N.S.), March 22, 1947.

7 Graham Metson interview with Charlie Murray, October 1979.

8 Interview with Alex "Scotty" Munro, Toronto, March 29, 1989; *New Maritimes*, November/December 1989, p.2.

9 *The Halifax Chronicle*, May 17, April 19, 1937.

10 *The Halifax Chronicle*, January 10, 1938.

11 *The Halifax Chronicle*, January 22, 1938.

12 Stephen Kimber, *Net Profits: The Story of National Sea* (Halifax: Nimbus Publishing, 1989), p.66.

13 The story of the Canadian Fishermen's Union is told in *The Lockeport Lockout*, researched by Michael Lynk, written by Sue Calhoun, published privately by Calhoun, Halifax, 1979.

14 *The Halifax Chronicle*, November 10, 1939.

15 *The Halifax Chronicle*, December 12, 1939.

16 William Repka and Kathleen Repka, *Dangerous Patriots: Canada's Unknown Prisoners of War* (Vancouver: New Star Books, 1982), pp.127–128.

17 Marguerite Woodworth, *History of the Dominion Atlantic Railway* (Kentville, N.S.: Kentville Publishing, 1936), pp.140–142.

18 Interview with Becky Dingee Fillmore, Wolfville, N.S., February 6, 1990.

19 *The Advertiser*, June 30, 1938; and interviews with family members.

20 *The Advertiser*, September 1, 1938.

21 *The Steelworker*, June 15, 1935.

22 *The Steelworker*, November 10, 1934.

23 Interview with Dane Parker, Toronto, September 21, 1988.

24 William Beeching and Phyllis Clarke, eds., *Yours In Struggle: Reminiscences of Tim Buck* (Toronto: NC Press, 1977), p.388.

FIFTEEN Party Politics and the War Against Fascism

1 *The Halifax Chronicle*, March 5, 1936.

2 Ibid. Roscoe's response to Gillis appeared in the same issue of the newspaper.

3 *The Clarion* (Toronto), August 12, 1939.

4 *The Clarion*, July 22, 1939; Lita-Rose Betcherman, *The Swastika and the Maple Leaf: Fascist Movements in Canada in the Thirties* (Toronto: Fitzhenry and Whiteside, 1975), pp.141–142.

5 *The Halifax Chronicle*, July 2, 7, 1936.

6 *The Halifax Chronicle*, March 10, 1937.

7 Gregory Kealey and Reg Whitaker, eds., *RCMP Security Bulletins: The War Series, 1939–41* (Toronto: Committee on Canadian Labour History, 1990), p.111.

8 Reg Whitaker, "Official Repression of Communism During World War II," *Labour/Le Travail*, Spring 1986.

9 *New Maritimes*, April 1987. Wallace was interned for eighteen months in Hull and

Petawawa. Shortly after his release in late 1942 his first collection of poetry, *Night Is Ended*, was published in Winnipeg. Five additional volumes of his poetry were published in Canada and the Soviet Union. Wallace's work, criticized for being politically motivated and bad verse, did not find acceptance in the mainstream of Canadian poetry. For several years before his death in 1975, Wallace wrote a column for *The Canadian Tribune*.

10 Repka and Repka, *Dangerous Patriots*, pp.128–129.

11 Ibid., p.129.

12 *The Varsity*, February 10, 1960, p.5.

13 Letter from the Minister of National Revenue, Ottawa, to Brig. S.T. Wood, Commissioner, RCMP, Ottawa, October 3, 1939.

14 Canadian Postal Censorship Report, Ottawa, June 10, 1941.

15 General Franz Halder, affidavit, November 22, 1945, at Nuremberg, quoted in William L. Shirer, *The Rise and Fall of the Third Reich: A History of Nazi Germany* (New York: Fawcett Crest, 1962), p.1088.

16 Interview with Kell Antoft, Halifax, October 12, 1990.

17 Interview with Irene Cunningham Fillmore, Halifax, September 27, 1990.

18 Telephone interview with Vern Bigelow, Vancouver, January 21, 1990. He said Roscoe first took him to the counsel offices on South Park Street in Halifax and introduced him to the Kutsenkos.

19 *The Halifax Herald*, December 10, 1943.

20 Ian Grey, *Stalin: Man of History* (Garden City, N.Y.: Doubleday, 1979), p.427.

21 Shirer, *Rise and Fall of the Third Reich*, pp.1242–1243.

22 *The Halifax Chronicle*, December 14, 1942.

23 S.W. Horrall, "Canada's Security Service: A Brief History," *RCMP Quarterly*, Summer 1985, p.45, cited in Kealey and Whitaker, *RCMP Security Bulletins*, p.10.

24 *The Advertiser*, January 14, 1943.

25 *The Advertiser*, April 8, 1943.

26 Tim Buck, *Thirty Years, 1922–1952: The Story of the Communist Movement in Canada* (Toronto: Progress Books, 1952), p.174.

27 Beeching and Clarke, *Yours In Struggle*, p.322.

28 Penner, *Canadian Communism*, p.194.

29 *Steelworker and Miner*, September 11, 1943.

30 *The Halifax Herald,* December 20, 1943. Among the people active in the party were: Chad MacMillan, Steve McPherson, Lawrence Carroll, Les or Leo MacIntosh of Glace Bay, Scott McLean, David Stewart, Alice Munro of Halifax, Murdock Clarke of Glace Bay, Earl Shaffleburg of Lunenburg, Vern Bigelow of Halifax, Patrick White of New Glasgow, Douglas Margeson of Halifax, and Eric Atkins of Halifax.

31 Telephone interview with Vern Bigelow, Vancouver, January 7, 1990.

32 Letter to the author from Fran Fassett, Vancouver, January 13, 1990.
33 RCMP report, Kentville, N.S., March 31, 1944.
34 Conrad, *George Nowlan*, p.77.
35 Interview with Irene Cunningham Fillmore, Halifax, September 27, 1990.
36 RCMP report, Commander of Halifax subdivision to RCMP Commissioner, Ottawa, January 3, 1944.
37 RCMP report, November 28, 1944.
38 From Fred Brodie interview for proposed National Film Board documentary, "The Centreville Socialists," 1978.
39 RCMP report prepared by Inspector F.T. Evens, Commander, Sydney, N.S., subdivision, December 19, 1944.
40 *Canadian Tribune*, December 18, 1943.
41 Conrad, *George Nowlan*, p.40.
42 *Steelworker and Miner*, February 26, 1944.
43 *Steelworker and Miner*, October 30, 1943.
44 RCMP report prepared by Cpl. E. Swailes, Kentville Detachment, August 17, 1944.
45 Roscoe Fillmore address on CFCY Radio, Charlottetown, March 13, 1944, cited in RCMP report, Halifax subdivision to RCMP Commissioner, Ottawa, March 21, 1944.
46 *Parliamentary Guide*, 1946.
47 Frank Fillmore letter to B. Boivin, Biosystematics Research Institute, Agriculture Canada, Ottawa, June 7, 1974.
48 *Steelworker and Miner*, November 2, 1946.

SIXTEEN Cold War Days: The Decline of the Labor-Progressives

1 John Sawatsky, *Men In the Shadows: The RCMP Security Service* (Toronto: Doubleday Canada, 1980), Chapter 6.
2 *Steelworker and Miner*, March 2, April 20, November 9, 1946.
3 Charlie Murray interview with John Bell and John Shuh, 1978, tape 18, Dalhousie University Archives, Halifax.
4 RCMP report, Halifax subdivision to RCMP Commissioner, Ottawa, January 28, 1946.
5 Telephone interview with Ethel Meade, Toronto, November 22, 1990.
6 RCMP report, Halifax subdivision to RCMP Commissioner, Ottawa, December 19, 1946.
7 E. Jean Nisbet, "'Free Enterprise at Its Best': The State, National Sea, and the Defeat of the Nova Scotia Fishermen, 1946–1947," in *Workers and the State in Twentieth Century Nova Scotia*, ed. Michael Earle (Fredericton: Acadiensis Press for the Gorsebrook Research Institute of Atlantic Canada Studies, 1989), p.171.
8 *Steelworker and Miner*, March 8, 1947.

9 *Steelworker and Miner*, May 22, 1947.
10 Kimber, *Net Profits*, p.120.
11 Green, *Against the Tide*, pp.225–227.
12 *Steelworker and Miner*, July 23, 1949.
13 *Steelworker and Miner*, May 15, 1948.
14 *The Halifax Mail*, March 15, 1948.
15 *The Halifax Herald*, May 13, 1948.
16 *Steelworker and Miner*, February 9, 1952.
17 Allan O'Brien letter to J.K. Bell, April 24, 1962, Dalhousie University Archives, Halifax.
18 Watson Kirkconnell papers, correspondence with RCMP, Acadia University, Wolfville, N.S.
19 Roscoe Fillmore letter to Ruth Fillmore Tarbuck, July 6, 1950.
20 *Steelworker and Miner*, April 10, 1948.
21 *Steelworker and Miner*, July 30, 1949.
22 *Steelworker and Miner*, January 8, 15, 1949.
23 *Steelworker and Miner*, April 12, 1947.
24 *Steelworker and Miner*, April 5, 1947.
25 *Steelworker and Miner*, April 12, 1947.
26 *Steelworker and Miner*, December 15, 1951.
27 *Steelworker and Miner*, January 29, 1949.
28 *Steelworker and Miner*, April 12, 1947, December 31, 1948.
29 Martin Luther King Jr. letter to Roscoe Fillmore, August 20, 1964, Dalhousie University Archives, Halifax.
30 *Steelworker and Miner*, March 31, 1951.
31 *Steelworker and Miner*, May 29, 1948, April 30, December 10, 1949.
32 *Steelworker and Miner*, May 26, 1951.
33 RCMP reports, Nova Scotia headquarters to RCMP Commissioner, Ottawa, May 19, June 23, 1951.

SEVENTEEN Mr. Green Thumbs

1 *The Dartmouth Free Press*, letter to the editor from Irene Cunningham Fillmore, September 10, 1975.
2 Letter to the author from Kenneth Wilson, Sardis, B.C., May 1, 1989.
3 Roscoe Fillmore, address to the International Plant Propagators Society, Cleveland, December 2, 1954.
4 Ibid.
5 Letter to the author from Kenneth Wilson, Sardis, B.C., May 1, 1989.
6 Letter to the author from Donald Craig, Kentville, N.S., January 5, 1989.
7 Fillmore, address to the International Plant Propagators Society.

8 Roscoe Fillmore, *Roses for Canadian Gardens* (Toronto: Ryerson Press, 1959), p.ix.
9 Mrs. D.E. Holland, Editorial Department, Macmillan of Canada Ltd., Toronto, letter to Roscoe Fillmore, December 30, 1952.
10 Frank Flemington, Assistant Editor, Ryerson Press, Toronto, letter to Roscoe Fillmore, December 30, 1952.
11 *Canadian Homes and Gardens*, October 1953.
12 *The Guelph Mercury*, January 9, 1946.
13 *The Advertiser*, July 7, 1977.
14 Telephone interview with Barbara Tarbuck, Los Angeles, January 28, 1990.
15 Roscoe Fillmore identical letter to Allie Fillmore Hoad, Belleville, Ont., and Ruth Fillmore Tarbuck, Detroit, August 5, 1954.
16 Ibid.
17 Roscoe Fillmore letter to Richard Fillmore, Boston, September 26, 1954.
18 Frank Fillmore letter to Richard Fillmore, Durham, N.C., May 27, 1957.
19 A.R. Buckley, Curator, Dominion Arboretum and Botanic Garden, Science Service Branch, Department of Agriculture, Ottawa, writing in *Canadian Geographical Journal*, February 1958.
20 Letter to the author from Donald Craig, Kentville, N.S., January 5, 1989.
21 Roscoe Fillmore letter to Richard Fillmore, Durham, N.C., February 9, 1958.
22 Victor Thorpe letter to George Nowlan, July 8, 1957, cited in Conrad, *George Nowlan*, p.175.
23 RCMP report, Kentville, N.S., June 2, 1954.
24 Cited in Kealey and Whitaker, *RCMP Security Bulletins*, pp.13–14.
25 Roscoe Fillmore letter to Becky Dingee Fillmore, Durham, N.C., November 10, 1956.
26 Roscoe Fillmore letter to Barbara Tarbuck, Detroit, September 12, 1968.
27 *The Canadian Tribune*, September 4, 1968.
28 Roscoe Fillmore letter to Richard Fillmore, Durham, N.C., March 10, 1958.

EIGHTEEN The Man Within

1 The Fillmores were one of seven families that received land grants in 1759–60 in what was known as Cumberland Township on Chignecto Isthmus. The seven thousand families that eventually arrived were known as the New England Planters and should not be confused with the United Empire Loyalists, who arrived later in the Maritimes. Esther Clark Wright, "Cumberland Township: A Focal Point of Early Settlement on the Bay of Fundy," in *They Planted Well: New England Planters in Maritime Canada*, ed. Margaret Conrad (Fredericton: Acadiensis Press, 1988). Conrad's article originally appeared in *Canadian Historical Review*, Vol.27 (1946), pp.27–32.

Charles L. Fillmore, in *So Soon Forgotten*, pp.27–29, explains that John Fillmore was a corporal in the U.S. military. He applied for a land grant in the colony of Nova Scotia in 1759, but he and his wife Leah may not have taken up residence there until sometime between 1759 and 1763, when a son was born in the colony. Shortly after their arrival the total population of that part of Nova Scotia, which later became New Brunswick, was barely one thousand people.

2 Roscoe Fillmore letter to Clara Fillmore Gross, Detroit, May 28, 1964.

3 Roscoe Fillmore letter to Dane Parker, Halifax, March 18, 1968.

4 Taped statement given the author by Mary Fillmore, Chestnut Hill, Mass., February 19, 1990.

5 Perly, "We Bury Our Poets"; Burris Devanney, "Shouting His Wares: The Politics and Poetry of Kenneth Leslie," *New Maritimes*, June 1986; Devanney, "Kenneth Leslie."

6 *The Canadian Tribune*, December 12, 1948; *Steelworker and Miner*, January 9, 1954.

7 Roscoe Fillmore letter to Dick and Becky Dingie Fillmore, Durham, N.C., December 28, 1956.

8 The name "Helen" is a pseudonym for the Saint John woman, who has since died.

9 Roscoe Fillmore letter to Herman Fillmore, Penobsquis, N.B., May 10, 1959.

10 Interview with Rosa Fillmore Skinner, New Minas, N.S., August 22, 1988.

11 Cyril Robinson and Bert Beaver, "180,000 Pansies," photostory in *Weekend Magazine* (Montreal), Vol.5, No.18 (1955).

12 Fillmore's Garden Centres Ltd. report to Nova Scotia Industrial Loan Board, Halifax, July 20, 1960.

13 Roscoe Fillmore letter to Nova Scotia Industrial Loan Board, Halifax, March 27, 1961.

14 *The Chronicle-Herald*, May 26, 1961.

15 Roscoe Fillmore letter to Dick and Becky Dingee Fillmore, October 21, 1966.

16 Roscoe Fillmore letter to Barbara Tarbuck, October 15, 1967.

17 Roscoe Fillmore letter to Dane Parker, Halifax, March 18, 1968.

18 Roscoe Fillmore letter to Dane Parker, Halifax, June 25, 1968.

19 Letter to Roscoe Fillmore from Tim Buck, Toronto, November 2, 1968, Dalhousie University Archives, Halifax.

20 Kenneth Leslie, *The Poems of Kenneth Leslie* (The Ladysmith Press, 1971), p.151.

Index

Abbreviations used in the index:
RF—Roscoe Fillmore
LPP—Labor-Progressive Party
DAR—Dominion Atlantic Railways
SP—Socialist Party of Canada
CPC—Communist Party of Canada

Printed and bound in Canada by
Best Gagné Book Manufacturers

818029